D1328395

Date Due

(

F

ADHESIVES FOR METALS
Theory and Technology

ADHESIVES FOR METALS
Theory and Technology

Nicholas J. DeLollis

Materials Engineer (Synthetic Resins Applications)
Sandia Laboratories, Albuquerque, New Mexico

INDUSTRIAL PRESS INC.
200 Madison Avenue, New York, N.Y. 10016

To Domenic and Zalmira
without whose cooperation I would not
be here to write this book.

668.3
D323a

Contents

List of Tables

LIST OF TABLES

Preface

Adhesion theories and practices are continually changing as new resin materials are developed and old materials are improved. With these new developments and improvements, we hope, comes greater understanding.

I have worked with synthetic resins and adhesives for about twenty-five years, and in that time I have been involved in some of the controversy associated with adhesives theory and practice. As my ideas gelled, I finally succumbed to the urge to put them down on the printed page. Since I am not mathematically inclined, I have purposely avoided symbolism in favor of reasoned discussion, hoping that my ideas will be helpful to engineers actually working with adhesives application.

I owe considerable thanks to the Sandia Corporation for allowing me to use data accumulated while working on Sandia developments. Furthermore, this book could not have been written without the cooperation of Sandia typing, editing, and drafting personnel. Many thanks are due, in particular to Jo, Marge, Kathy, and Della, who learned frustration tolerance in the process of deciphering my hieroglyphics; to Edward Dlouhy who did many of the drawings; and to Lee Class who put the final polish on what I hope will be a gem. Finally, many thanks to my friends at Sandia, who, with their diversified scientific skills, helped through our discussions, to keep my thoughts well defined and uncluttered by biased ideas and wishful conclusions.

CHAPTER 1

Introduction

THE ORIGINS AND HISTORY OF ADHESIVES [1,2]

Man has used adhesives for many centuries. At first he simply learned to use materials found in nature. Mongol bowmen, for example, used adhesives probably made from the hide, hooves, bones, and even the blood of horses in making their laminated-wood bows. But man has always had the urge to imitate, influence, and duplicate nature. Among the first to follow this urge were imaginative, enterprising people known as alchemists, witches, wizards, and magicians. Many did little but inspire some wonderful stories—but some undoubtedly were individuals with genuinely inquiring minds who were not satisfied to let Mother Nature remain unchallenged and were constantly trying to outdo her.

Over the centuries, the old materials were refined and developed to the peak of their utility, but major changes were slow to occur. Some violin makers and cabinet makers of today are still using adhesives that were used in ancient Egypt. To a certain extent, the understanding of the old materials and processes became proprietary knowledge handed down from father to son, with the result that newcomers, finding it difficult to duplicate the ancient arts, were forced to turn to new materials.

The emergence of adhesives technology as a factor in industry took place in Europe in the eighteenth century and in the United States in the nineteenth century. At that time, adhesives were obtained almost entirely from animal and vegetable materials such as hide, hoof, bone, and blood proteins and from tree gums, casein, or starches.

The synthetic resins of today are made up chiefly of atoms of carbon, hydrogen, oxygen, and nitrogen, which are present in the air and in organic matter in virtually unlimited quantities. Silicon and sulfur, both common elements, are used in the silicone resins and the polysulfides, respectively. Other well-known elements such as copper, tin, lead, and titanium are used as catalytic ingredients.

SYNTHETIC RESINS

Synthetics saw the light of day early in the twentieth century with Dr. Baekeland's development of phenolic resins. The phenolic resins—not only the earliest but also the most prolific in terms of offspring developments—proved almost indispensable as ingredients in metal adhesives and in high temperature adhesives and resins for use in aerospace structures.

By itself, phenol formaldehyde is not suitable for use with metals. Not only is it too hard and brittle, but the water that is a by-product of the polymerization reaction interferes with resin adhesion to impermeable adherends such as metals. Phenol formaldehyde was first successfully applied to wood. Wood is a satisfactory adherend in this application; having a low modulus, at least in one direction, it can adapt itself to the hard, brittle resin, and its porous structure facilitates dissipation of the water by-product.

In spite of their original limitations, the phenolic resins proved versatile when they were co-polymerized with various rubbers, acetals, and epoxies, and the age of structural metal adhesives was well under way.

SEALANTS

Materials of another class play an important role in metal fabrication; these are sealants, whose development parallels that of the structural adhesives. The sealants are not normally considered as adhesives, yet in order to function properly they must adhere as well as or better than the structural adhesives, because they must withstand many types of conditions ranging from continuous immersion in all sorts of liquids to exposure to aerospace conditions and heat of atmospheric reentry. Under such conditions the sealants may be used either by themselves or as a protection for the structural adhesive (for example, edge sealants for bonded aircraft panels).

At one time, sealants were not required to exhibit great strength, but this is no longer true. Recent developments in polyurethane castable rubbers and silicone sealants have resulted in materials with tensile, peel, and tear strengths which equal and sometimes exceed those of the so-called structural adhesives. Thus under the proper conditions these materials can also be used in structural applications, especially where vibration damping and stress relief are required.

The polysulfide rubber sealants were the first to achieve wide use. Because of their resistance to petroleum fuels, they were first used as fuel tank sealants in aircraft, but have since found wider application in the automobile and building industries. During a search for information concerning the long-term aging properties of polysulfides, I learned from the Douglas Aircraft Company that the history of polysulfide sealants parallels that of the DC-3 aeroplane, in that polysulfide sealants were used in its fuel tanks. First built in 1936, the DC-3 is still in use all over the world.

SERVICE LIFE

The history of the polysulfide sealants points out a universal problem in the field of synthetic resins: The basic polymers and subsequent commercial formulations are continuously in the process of being changed with a view to improvement. Thus the polysulfide formulation which was used in the first DC-3 is no longer in existence. As new and better formulations and specifications are developed, the old are discarded. The result is that a manufacturer may want to make a reliable, long-lasting product, resistant to many environmental conditions, but finds himself handicapped by a lack of long-term aging data on present formulations, which are supposedly superior to older ones. To a certain extent the manufacturer must proceed on faith, and hope that time will resolve the dilemma in his favor.

Unfortunately, the task of preparing a program to determine the effects of aging on a given material—where one must minimize the number of variables in order to maintain a workable plan and at the same time retain enough significant variables so that the results are worthwhile—becomes so involved that many a plan dies from the inertia which increases in proportion to the plan's magnitude.

ADHESIVE/ADHEREND COMPLEX

The property of adhesives and sealants that makes them unique is that they have no identity by themselves. They are adhesives and sealants only to the extent that they bond to or seal some other material or materials.

The usual materials specialist is concerned with the bulk properties of a specific material; that is, modulus, temperature coefficients, density, and so on. His concern with surface properties is usually limited to those which affect the bulk properties, for example, crazing, corrosion initiation, stress cracking fracture, permeability. Thus his field of interest is essentially an isolated entity.

Materials Engineer

The adhesive materials engineer, however, faces a much more complex situation. He may be dealing with a material which is available in the uncured state as a liquid or without solvent; a solid which must be heated to a liquid before being used; or a solid film which requires heat and pressure to achieve the fluid state necessary for successful application. In each case, he must know and correctly specify the time, temperature, pressure, and viscosity relationship and must evaluate the ability of any one of these materials to wet and properly attach itself to a wide variety of adherends. Also, he must determine the proper surface treatment of the adherend necessary for achieving wettability and subsequent acceptable level of bond strength. The bulk properties of the adherends must be known, since if these are not similar they must be analyzed together with the properties of the adhesive material, to insure that the resulting stresses do not far exceed any possible bond strength or adherend strength. When all of these problems have been resolved, the adhesive materials engineer still has to be concerned with the physical and chemical stability of the bonded assembly with relation to its intended environment.

Materials (such as Teflon®) which were until recently considered unbondable, eventually yielded to the persistent efforts of the laboratory worker and today are easily bonded. Adhesives which at one time were limited to use in protected internal environments with limited temperature fluctuations have now been improved to the extent that they can successfully withstand temperature extremes ranging from −240C (−400F) to 260C (500F) operationally and to 816C (1500F) for limited periods of time. Resistance to aqueous and organic solvents has been similarly improved.

Advanced Development

A measure of the success with which synthetic adhesives and sealants have kept up with the times is that when the United States agreed to go ahead with the construction of the supersonic transport with its 260C (500F) skin temperatures, structural adhesives based on polyimide and polybenzimidazole resins were already available with thousands of hours of evaluation at 260C (500F). Fluorinated silicone materials had already been developed as fuel-resistant high-temperature sealants, and room temperature curing (RTV) silicone sealants and adhesives were ready for use at temperatures up to 316C (600F).

The future of these materials is particularly bright when one considers that the synthetic resin industry is only about thirty years old and seems to be accelerating its output rather than leveling out at a plateau. Recent developments such as polysulfones, polyphenylene oxides, borosilanes, polyquinoxylenes, and others emphasize the fact that the possible combinations and permutations of molecular structures are practically unlimited. Future possibilities are further enhanced by a synthetic resins industry which has the imagination to anticipate future requirements and the faith to invest in the work necessary to achieve them.

ADVANTAGES AND DISADVANTAGES OF ADHESIVE BONDS

Advantages:

The chief advantages of an adhesive bonded structure may be summarized as follows:

1. Possibly, the principal advantage of an adhesive bond is that the adhesive fastens to the entire bonded surface, thereby distributing the load more or less evenly (depending on the modulus of the adhesive) and thus avoiding highly localized stress. A nail driven into wood may crack the wood; a rivet in thin-gauge metal may initiate fatigue cracking. Use of adhesives overcomes these problems.

2. In the manufacture of aircraft and automobiles, adhesive bonding may be used to advantage instead of riveting and welding, since (a) riveted structures add weight and drag, (b) rivets and welds are unsightly and difficult to conceal, and (c) holes and welds may facilitate the start of corrosion. Adhesives in film form are particularly desirable for this type of application, since they facilitate close control of weight distribution and total weight of adhesive used.

3. In the presence of vibration, a bonded structure usually has a longer life than a riveted assembly. In addition, because adhesives of different moduli are available, damaging resonant frequencies can be modified or even eliminated by proper analysis and selection of the adhesive.

4. A bonded joint is a sealed joint. Aircraft wing-tanks and aluminum, honeycomb panels are sealed in this manner. This saves
   considerable weight and simplifies construction.
5. Adhesives are usually electrical insulating materials. When
   different metals are bonded together, this minimizes the possibility of electrolytic corrosion. When positive separation in the
   form of glass fiber cloth is incorporated in the bond, an electrically insulating layer becomes a part of the bonded assembly.
   Conversely, silver-filled adhesives make possible electrically
   conductive bonds between adherends.
6. Adherends with different coefficients of expansion can be
   bonded with low modulus or rubbery adhesives. Glass-to-metal
   and ceramic-to-metal joints can be made without stressing
   the glass or the ceramic. This makes it possible to bond ablative shields to missile structural shells. Otherwise it would be
   necessary to devise elaborate mechanical fasteners which
   would allow the shield to expand independently of the case
   and which would not conduct external heat to the structural
   case.
7. Adhesive sealants can be cured at comparatively low temperatures and do not melt on subsequent reheating. They are often
   used in preference to soldering, brazing, or welding since the
   high temperatures associated with the latter processes sometimes warp structures, crack glass-to-metal seals, or damage
   heat sensitive components.

Disadvantages:

1. The main disadvantage is that an adhesive bond, unlike a
   riveted joint, does not permit visual examination of the bond
   area. The continuity or lack of continuity in the bond cannot
   be seen, and evaluation must be destructive. Nondestructive
   test methods such as the various ultrasonic techniques can indicate only a complete lack of bonding, not the degree of bonding. It is sometimes possible to test bonded items to a certain
   percentage of ultimate strength if the size and shape are convenient.
2. Surface cleanliness and good process control are important in
   any bonding process. This requires considerable equipment,
   depending on the critical nature of the application.
3. Holding fixtures, presses, ovens and autoclaves, not usually

needed for other fastening methods, are necessities for adhesive bonding.

4. At present, adhesives cannot be used where extended life above 316C(600F) is a requirement.

TEMPERATURE DESIGNATION FOR ADHESIVE BONDING

The trend toward the centigrade–metric system is gaining headway in the United States. The American Society for Testing and Materials (ASTM) has issued recommendations for conversion to the metric system and for manner of presentation. These recommendations are contained in the manual, *Recommendation on Form of ASTM Standards*, January 1964, and in the pamphlet, *Recommended Procedures for Metric Conversion in ASTM Standards*, March 1966.

As international travel, communication, and trade increase and the precarious goal of worldwide unity comes nearer to reality, metric unification becomes inevitable and should be encouraged. Accordingly, in this book temperatures will be expressed in centigrade or Celsius units with Fahrenheit degrees in parentheses. However, since it would not be useful to give the exact Fahrenheit equivalent to the nearest decimal, the Fahrenheit temperatures will be rounded off to the nearest whole degree. Temperature ranges will be similarly treated.

NOMENCLATURE

Like other industries, the adhesives industry has developed its own nomenclature. ASTM defines about two hundred terms in the 1967 *Book of Standards*.

A few of the more commonly used terms are listed here:

Adherend—A body which is held to another body by an adhesive.

Adhesion—The holding together of two surfaces by interfacial forces which may consist of valence forces or interlocking action, or both.

Adhesive—A substance capable of holding materials together by surface attachment.

Assembly—A group of materials or parts, including adhesive, which has been placed together for bonding or which has been bonded together.

Binder—A component of an adhesive composition which is primarily responsible for the adhesive forces holding two bodies together.

Bond strength—The unit load applied in tension, compression, flexure, peel, impact, cleavage, or shear required to break an adhesive assembly with failure occurring in or near the plane of the bond.

Catalyst—A substance which markedly speeds up the cure of an adhesive when added in minor quantity as compared to the amounts of the primary reactants.

Consistency—That property of a liquid adhesive by virtue of which it tends to resist deformation.

Creep—The dimensional change with time of a material under load, following the initial instantaneous elastic or rapid deformation. Creep at room temperature is something called *cold flow.*

Diluent—An ingredient usually added to an adhesive to reduce the concentration of bonding materials.

Elastomer—A material which at room temperature can be stretched repeatedly to at least twice its original length and, upon immediate release of the stress, will return with force to its approximate original length.

Flow—Movement of an adhesive during the bonding process, before the adhesive is set.

Failure, adhesive—Rupture of an adhesive bond, such that the plane of separation appears to be at the adhesive-adherend interface.

Failure, cohesive—Rupture of an adhesive bond, such that the separation appears to be within the adhesive.

Film, adhesive-supported—An adhesive supplied in sheet or film form with an incorporated carrier that remains in the bond when the adhesive is applied and used.

Film, adhesive-unsupported—An adhesive supplied in sheet or film form without an incorporated carrier.

Hardener—(1) A substance or mixture of substances added to an adhesive to promote or control the curing reaction by taking part in it. (2) A substance added to control the degree of hardness of the cured film.

Inhibitor—A substance which slows down chemical reaction, sometimes used in certain types of adhesives to prolong storage or working life.

Joint—The location at which two adherends are held together with a layer of adhesive.

Joint, starved—A joint which has an insufficient amount of adhesive to produce a satisfactory bond.

Modifier—Any chemically inert ingredient added to an adhesive formulation that changes its properties.

Permanence—The resistance of an adhesive bond to deteriorating influences.

Plasticizer—A material incorporated in an adhesive to increase its flexibility, workability, or distensibility. The addition of the plasticizer may cause a reduction in melt viscosity, lower the temperature of the second-order transition, or lower the elastic modulus of the solidified adhesive.

Primer—A coating applied to a surface, prior to the application of an adhesive, to improve the performance of the bond.

Shelf life—The period of time during which a packaged adhesive can be stored under specified temperature conditions and remain suitable for use.

Storage life—Same as shelf life.

Temperature, curing—The temperature to which an adhesive or an assembly is subjected to cure the adhesive.

Temperature, drying—The temperature to which an adhesive on an adherend or in an assembly, or the assembly itself, is subjected to dry the adhesive.

Thermoplastic—A material which will repeatedly soften when heated and harden when cooled.

Thermosetting—Having the property of undergoing a chemical reaction by the action of heat, catalysts, ultraviolet light, etc., leading to a relatively infusible state.

Viscosity—The ratio of the shear-stress existing between laminae of moving fluid and the rate-of-shear between these laminae.

REFERENCES

1. Delmonte, J. *The Technology of Adhesives.* New York: Reinhold Publishing Corporation, 1947.
2. Skeist, I. *Handbook of Adhesives.* New York: Reinhold Publishing Corporation, 1962.

CHAPTER 2

Theory of Adhesion and Mechanism for Bond Failure[*]

INTRODUCTION

In spite of the advances made in synthetic resin development, the problem of developing a new adhesive or in improving an old formulation is usually approached empirically. The formulator, on the basis of past experience, lays out a program in which he will juggle an infinite variety of vulcanizing agents, inorganic fillers, reinforcing fillers, antioxidants, plasticizers, accelerators, retardants, wetting agents, organic solvents, emulsifiers, pigments, and resin modifiers. The adhesives will be tested and evaluated on the basis of resistance to temperature, atmospheric environment, organic solvents, etc. Those which successfully pass the tests will become stock items.

We have stated that the development of an adhesive and adhesion is often the result of experience coupled with trial-and-error. But that is not the end of the process. Samplings by manufacturers may reveal flaws, and this will trigger a hasty withdrawal of the product from the market. The adhesive then goes back to the laboratory for further study and modification before being marketed again, possibly with fewer claims in its favor. One such premature adhesive presentation occurred when polyvinyl-acetate emulsions were offered as the ideal furniture adhesives. They were quickly withdrawn when it was found that the cold-flow properties had not been properly evaluated.

THEORIES

In this section, a brief description of the evolution and present status of the theories of adhesion will be followed by (1) a proposed

*Portions of this chapter were published in *Adhesives Age*, December 1968–January 1969.

mechanism of bond failure which is intended to connect theoretical and experimental knowledge of bond strengths by showing reasons for failure, and (2) a practical guide for the development of optimum bond strengths, especially in the area of structural-metal adhesive designs. It is hoped that this will result in a more scientific and less empirical approach to the development of adhesive systems and of structural bonded assemblies.

Bond Requirements—Cleanliness

It is generally recognized that in order to form an effective adhesive bond, one must start with a clean surface, i.e., a surface from which foreign materials such as grease, dust, dirt, liquids, and loose oxides have been removed. The surface will then be at equilibrium with atmospheric contaminants, which means that the principal contaminants are probably gases and water-vapor. Studies by White[1] have shown that some freshly prepared surfaces achieve

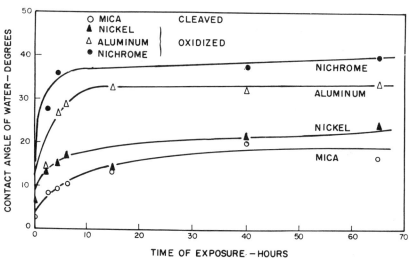

Fig. 2-1.   Contamination of surfaces exposed to laboratory air.

equilibrium with laboratory atmospheric contaminants in about five hours. (Fig. 2-1).

Fluidity

The adhesive in fluid form must be able to spread over the adherend surface, either spontaneously or with heat and pressure.

The greater the fluidity and the degree of wetting, the better will be the ultimate bond.

Solidification

The next process is the solidifying of the adhesive, which includes the following steps: (1) chemical polymerization either to a linear, thermoplastic solid or to a three-dimensional cross-linked solid, with or without the use of heat and pressure; (2) physical cooling from a liquid hot-melt, with or without pressure, to a solid; and (3) drying due to solvent evaporation. The final state is that of a high-strength solid strongly attached to the adherend.

The forces which cause the adhesive to wet, spread, and attach the solidified liquid to the adherend have been ascribed to chemical bonds, to mechanical entanglement, to physical and chemical adsorption due to polar groups, to electrostatic forces of attraction inherent in all matter, and to combinations of these factors. A more detailed description of these various theories follows.

Chemical Bond Theory

One of the first theories (the Chemical Bond theory of adhesion)[2] related adhesion between adhesive and adherend to the forces holding atoms and molecules together. These forces, as described by Bikerman,[3] are 330,000 psi for electrostatic forces acting between ions for crystals such as NaCl. For polymers such as phenolformaldehyde, where the breaking of primary valence bonds is involved, the attractive force may be 6,000,000 psi. These astronomical figures represent idealized stress-free structures which do not exist in nature, but they emphasize the gap between theory and reality. In actuality the best of bonded assemblies may result in strengths of 10,000 psi, which may reflect the limiting cohesive-strength of the adhesive or adherend. Other calculations based on the theoretical shear-strength, or the energy necessary to move molecules across a surface, achieve a much lower, but still unrealistic, figure of 56,000 psi.[4] Nonetheless, if we look closely at the structures involved in bonding applications where primers are used, we find that the chemical bond still has a place in the theory of adhesion.

Because the success of primers as adhesion promoters for resin glass-fiber systems and for RTV (room temperature vulcanizing) silicones has been well-documented,[5-7] and since these primers have chemically active terminations which must of necessity take part in the bonding process, we must conclude that chemically-active primers improve adhesion, and that the chemical bond theory is valid and is definitely making a place for itself in the general picture of adhesion theory.

Some of the organic molecules responsible for this improvement in adhesion (Table 2-1) are chemically-active materials similar to the ethoxy silanes, with the following structural formulas:[5]

$$NH_2CH_2CH_2CH_2Si(OC_2H_5)_3$$
Gamma-amino propyltriethoxysilane (A-1100)

$$CH_2 = CHSi(OCH_2CH_2OCH_3)_3$$
Vinyl-tris (b-methoxyethoxy) silane (A-172)

Equally significant improvement in bond strength is realized with the use of primers for bonding of RTV silicones to a wide variety of substrates. Without the primers, shear and T-peel strengths are negligible and would make the RTV resins virtually useless as adhesives and sealants. With the primers, shear strengths of 300 to 400 psi and T-peel strengths of 80 psi, with 100 percent cohesive failure, are not unusual. Unfortunately, since these materials have been developed in the laboratories of private industry, the exact materials used and the theories from which they were developed are protected as proprietary information.

Table 2-1. Flexural Strengths of Glass-Cloth-Reinforced Laminates

Laminate		Dry	After 2-hr Boil	Retention, percent
Resin	Silane	Flexural Strength, psi		
Polyester	A-172	72,000	68,000	94
	A-1100	37,000	18,000	48
	None*	60,000	35,000	58
Melamine	A-172	31,500	28,000	89
	A-1100	90,000	85,000	94
	None*	38,000	27,000	71
Epoxy	A-172	75,600	70,000	93
	A-1100	81,000	75,000	93
	None*	73,600	53,500	75
Phenolic	A-172	60,000	18,000	30
	A-1100	80,000	55,000	70
	None*	69,000	14,000	20

* Heat-cleaned glass cloth.

The actual chemical nature of the improved adhesive bond resulting from the use of primers has not been spelled out in detail. However, the results do not comply with the weak boundary layer
theory;[8] nor do they accord with the surface energy theory[9] to the
extent that proponents of the surface energy theory claim that a
well-prepared bond cannot fail in adhesion.[10]

Weak Boundary Layer Theory

According to the weak boundary layer theory, as first described
by Bikerman[8] and somewhat later by Schonhorn et al.,[9] poor adhesion or adhesive failure in a bond is due to an inclusion of low-
molecular-weight liquid, or possibly, an unpolymerized liquid-
fraction of the adhesive present at the adhesive-adherend interface.
This liquid forms a weak boundary layer which prevents the proper
bonding.

Various reasons for the beneficial results of primers have been
advanced. In the case of the multifunctional silanes, results have
been partly explained by the formation of tough fracture-resistant
skins on the glass fibers or the creation of a water-repellent interface
between the resin and the glass. In the case of the RTV systems,
one can only suspect that the bonds may be chemical. The primers
are relatively small molecules and in the absence of chemical linking
they would fit the description of a weak boundary layer; instead,
they significantly improve adhesion. Thus, we must conclude that
the silanes are tightly bound in place rather than loosely held as
an intermediate or weak boundary layer. It is generally believed
that, because of the magnitude of the available attractive forces,
adhesive failure is not possible in a properly-prepared bond.[3,4] However, because the resin-glass bonds weaken after exposure to an
aqueous environment, and since the strength in an aqueous environment improves after a glass-surface treatment, we must assume that
the previous failure was in adhesion and that the primers were
developed on the theory that it was adhesion which had to be
improved. Subsequent significant improvement in bond strength
enhances the initial assumption that bond failures could be blamed
on failure in adhesion.

In the case of the RTV silicones, which achieve little or no
adhesion without primers, we could assume the presence of a weak
boundary layer material, since weight-loss determinations show that
RTV silicones contain up to 5 percent low-molecular-weight
silanes.[11] The use of a very thin layer of primer with a room-

temperature cure does not remove the volatile matter but does improve adhesion to the extent that failures are grossly cohesive in the RTV.

Surface Energy Theory

According to the surface energy theory, a liquid, in order to spread on a solid surface, should have a lower surface tension than the solid's critical surface tension. In terms of surface energetics if we assumed that, because of lack of adhesion due to surface contamination, the surface tension of the liquid was greater than the critical surface-tension of the contaminated solid, that is, $\gamma(LV) > \gamma_c$,[12] then an intermediate layer, which functioned only on the basis of surface-energetics criteria, could not possibly help. If the surface tension of the primer or intermediate layer were made less than the critical surface tension of the solid in order to wet and bond to the solid surface, the surface tension would then be considerably less than that of the RTV. The RTV (which, on the basis of this theory, has too high a surface tension to wet the original solid substrate) would then have much too high a surface tension to wet and bond to the primer or intermediate layer. Thus, some other factor, such as chemisorption, must be responsible for this type of adhesion.

Inherent Roughness Theory

A theory of adhesion based on the inherent roughness of surfaces is advanced by J. J. Bikerman.[13] While the ability of the adhesive to wet and spread over an adherend surface is recognized as important, once this has been achieved the mechanical strength of the cured resin coupled to the rough surface is emphasized as the basic reason for the strength of the adhesive bond.

Since this surface coupling is limited by such factors as shrinkage stresses, trapped gas-bubbles,[14] imperfect molecular-fit,[15] etc., the effect of surface roughness would be more important in shear, than in direct tension. The principal effect of surface roughness would then be that electrostatic forces would have greater surface area in which to operate. Eley estimates that attractive forces due to physical adsorption perpendicular to the surface of the adherend are about four times the forces necessary to move molecules parallel to the adherend surface.[4] The magnitude of these forces is more than enough to account for experimental values. It should be sufficient to say that the effect of roughness would be additive but not exclusive, that is, adhesive strength in shear would be the sum of electrostatic forces in shear plus the effect of roughness on shear strength.

Polar Theory

An earlier theory, as initially proposed by DeBruyne, states that polar adhesives would bond to polar adherends and that nonpolar adhesives would bond to nonpolar adherends.[16] This hypothesis, which says, essentially, that materials must be similar in nature to attract and bond to each other, governed some of the earlier work on adhesive formulation.[17] However, the theory has undergone extensive modification as surface-tension studies have thrown more light on wetting phenomena.

The thermodynamics of wetting states essentially, that if the work of adhesion of the liquid for the solid is greater than the work of cohesion or surface tension of the liquid, then the liquid will wet and spread on the solid. Thus, to achieve spreading, it is necessary that the critical surface-tension of the solid be equal to or greater than the surface tension of the liquid, as: $\gamma_c \geq \gamma LV$.

Zisman and others describe the solid-liquid-vapor relations in detail and show that adhesion is dependent on the thermodynamics of wetting and spreading.[12,18]

Organic liquids and water have surface-free energies of less than 100 ergs/cm^2 while metals, metal oxides, and metal salts have greater surface-free energies[19] ranging from 100 to 3000 ergs/cm^2. Organic liquids and water should, then, spread freely on clean, solid surfaces protected from atmospheric contamination.

The foregoing explains why organic thermosetting resins and thermoplastic hot-melts wet and bond to metals;[20] and, to a certain extent, it also explains adhesion of organic resins to other organic resins of greater surface energies.[21]

Attractive Forces

The attractive forces involved in the wetting-and-spreading phenomena, as evaluated by Zisman, are considered to be primarily London dispersion-forces modified by the presence of polar groups which contribute dipole-ionic and hydrogen-bonding interactions.[18] Fowkes has calculated the energy of these polar groups to be in excess of dispersion forces for water/organic-liquid interactions.[22] Dispersion forces are the major factor for such systems even in the presence of hydrogen-bonding groups.

Inconsistencies in the Theories

The foregoing presents, briefly, the current status of the various theories of adhesion; we will now attempt to point out some of their shortcomings.

The adhesive strength to be derived from the theoretical forces of attraction, as calculated by the theoretician, exceeds reality by a wide margin.[3,4] While there have been attempts to bridge this gap by showing that interfacial stresses due to shrinkage, air entrapment,[8,13,14] and high adhesive modulus tend to reduce the bond strength, it has been assumed that failures in properly-prepared bonds must, of necessity, be cohesive either in the adhesive or adherend because of the great disparity between the theoretically derived forces of attraction and the adherend strengths.

Surface Studies

Fortunately, workers in the field of adhesives and adhesion have been sufficiently skeptical to undertake corrective research on the assumption that observed failures were, in fact, adhesive, that is, in adhesion. Subsequent improvement in bond strength vindicated their assumption that adhesion failure is a significant factor in bond failure and corrective action must include the study and modification of surfaces.

Modification of surfaces has resulted in significantly-improved adhesive bonds with resin/glass-fiber systems,[23-26] bonding with polyurethane resins, and bonding with RTV silicones. These three systems will be described in greater detail under "Means of Bond Improvement," in Chapter 3.

Another area of inconsistency is the proposal that weak boundary layers, as described under "Theories," are responsible for seemingly adhesive failures. Obviously, bonds can be prepared on dirty surfaces and must fail in a weak interfacial layer; therefore, we will confine our discussion to well-prepared surfaces.

Commercial polyethylene is described as a compound containing weak boundary-layer material; however, when used as a hot melt, it can produce bonds which fail cohesively and which are limited only by the cohesive strength of the polyethylene. Surface treatments—either chemical (for example, sulfuric acid, dichromate solution), electronic,[27] or morphological[28]—can be used to significantly improve the bond strength of polyethylene. If it is assumed that these treatments improve adhesion by removing a weak boundary layer, then the improvement should be temporary, since none of the treatments has created an impermeable barrier. In polyethylene, the weak boundary layer material is usually assumed to be uniformly distributed throughout the compound, so that when some of it is removed from the surface, the remainder tends to permeate to the surface and reestablish the weak boundary layer. Henderson indicates that exudation would re-establish the surface layer within 50 hours.[29] The treatments mentioned above certainly accomplish the object of improving adhesion, but the connection between this accomplishment and weak boundary layers is rather tenuous.

Industrial Practice

Conversely, industrially-proven use of RTV silicones and poly-sulfides[30] as adhesives, and bonds to plasticized polyvinyl chloride with nitrile rubber[31] and polyurethane[32] adhesives has shown that good bonds are possible where compatible liquid inclusions are known to exist, either as low-molecular-weight fractions or as plasticizers. Thus, weak boundary layers need not be significant factors in adhesion.

It is evident from the foregoing discussion that, to a certain extent, the present theories of adhesion are too confining and narrow in their coverage. What is needed is a theory which has greater perspective to provide a more realistic understanding of existing phenomena.

PRESENT TREND IN RESEARCH

Some idea of the present trend in adhesion research can be gained by a brief survey of the papers presented at the Chicago meeting of the ACS Division of Organic Coatings and Plastics Chemistry in September 1967. In the first section on interactions of liquids at solid substrates, four papers were presented describing the use and function of chemically active organic compounds as adhesion promoters, and one, already mentioned, on the effect of morphology on bondability of polyethylene.[28]

The first paper is concerned primarily with the preparation of improved adhesion promoters, with emphasis on the presence of a terminal group which could bond chemically to a solid substrate.[33] In the second paper, Zisman discusses the need for chemically active coupling agents with resistance to hydrolysis as a property necessary for long time durability.[34] Data is also presented to show that these coupling agents are wet better by organic liquids than by water, when applied on a metal surface. The paper by Bascom describes some work on the use of silane coupling agents on silica, to improve resin adhesion.[35] The fourth paper discusses the role of polymeric films of titanium dioxide as adhesion promoters on iron powder, E-glass, and silica.[36] The results show that the bonds to silica and E-glass are chemical in nature.

These papers indicate that adhesion promoters are being given increasing attention. Evidence is continually being accumulated as to the importance of the chemical bond in adhesion.

MECHANISM FOR BOND FAILURE

Having discussed the various theories of adhesion, we will now take a look at a possible mechanism for adhesive failure and see how this ties in with the theories.

Contact-Angle Shortcomings

The contact angle of water on metal has been widely used as a measure of cleanliness and bondability. However, on clean high-energy surfaces, all organic liquids and water should spread equally well to give a zero contact angle;[18] therefore, the contact angle cannot be used to discover preferential affinities of liquids for metals. Heats of wetting provide a far better measure of affinity or work of adhesion of various liquids for metals and metal oxides. Table 2-2 provides such a measure. These data, collected by Zisman,[12] from the literature, show that water has a greater affinity for polar and nonpolar solids than any of the other liquids tested, by a factor of almost three.

It has been shown that as little as 0.02 percent of water in benzene will give almost the same heat of wetting on a polar solid as pure water.[40] Eley cautions that polar impurities, especially water, in heats of wetting experiments will preferentially wet the surface and give spurious heats.[4]

The literature also indicates that once a liquid other than water is adsorbed on a solid, it can be preferentially desorbed by water or by any liquid which develops a greater work of adhesion for that particular solid.[1] All that is required is time for the water or other desorbing liquid to permeate the contaminating layer. Thus, an initial, large contact angle can be misleading.

Table 2-2. Literature Values of $f_{sv}o$ for Nonmetallic High-Energy Surfaces[a] [erg/sq cm at 25C (95F)]

Solid	Liquid	$f_{sv}o$ [d] $\equiv \gamma_s o - \gamma_{sv}o$	W_A (work of adhesion)[e]
		erg/sq cm @ 25C (95F)	
TiO_2	Water	300 (196[b])	370 (340[b])
	1-Propanol	114 (108[b])	138 (154[b])
	Benzene	85	114
	n-Heptane	58 (46[b])	78 (86[b])
SiO_2	Water	316	388
	1-Propanol	134	158
	Acetone	109	133
	Benzene	81	110
	n-Heptane	59	79
$BaSO_4$	Water	318	390
	1-Propanol	101	125
	n-Heptane	58	78
Fe_2O_3	n-Heptane[c]	54	94
SnO_2	Water	292 (220[b])	364 (364[b])
	1-Propanol	104 (117[b])	128 (163[b])
	Propyl acetate[b]	104	151
	n-Heptane[c]	54	94
Graphite	Water	64	136
	1-Propanol	95	118
	Benzene	76	96
	n-Heptane	57[c]	97[c]

[a] Data from Reference 37 unless otherwise indicated.
[b] Reference 38.
[c] Reference 39.
[d] $f_{sv}o$ = surface free-energy decrease, or surface-tension decrease on immersion of the solid in the saturated vapor of the liquid.
  $\gamma_s o$ = surface free-energy of solid or surface tension.
  $\gamma_{sv}o$ = surface free-energy of the solid immersed in the saturated vapor of the liquid.
[e] $W_A - W_{A'} = f_{sv}o$
  $W_{A'}$ = work to pull liquid away from surface leaving equilibrium-adsorbed film.

The phenomenon of adsorption and preferential displacement is widely applied in gas and liquid chromatography and is not restricted to low molecular weight materials. Separation of polymers with molecular weights as high as 2,000,000 is described by Smith.[41] Given sufficient time, desorption is not limited by molecular weight.

Stress Development

Since all the work of adhesion is accomplished in the initial wetting, spreading, and adsorption of the fluid on the solid surface, subsequent solidification in the case of adhesive liquids cannot increase the attractive forces. On the contrary—the stresses induced during solidification can only weaken the bond.[14,18] These stresses are greatest at the adhesive-adherend interface and are caused by (1) shrinkage in the adhesive during cure due to solvent loss, to chemical polymerization, to cooling, or to combinations of all three; (2) increase in modulus as the adhesive hardens; (3) gas bubbles trapped at the adherend surface, thereby limiting coverage and creating points of localized stress.

Work has been done to show that liquid/liquid/solid systems are not necessarily static,[42,43]—that is, one liquid will tend to displace the other at the metal surface depending on the relative affinities.[1] In a water-benzene-carbon system an equilibrium displacement pressure of 400 gm/cm^2 has been measured—the pressure necessary to halt the action of benzene in displacing water at the carbon surface. However, area tends to be meaningless when one realizes that displacement forces are acting along a front so that localized displacement stresses could be considerably greater than any measured area stress.

Observation of the displacement of oil or air by water from a hydrophilic surface shows that the oil or air globules at the surface are displaced along a curvilinear front. The localized stresses along this kind of front could assume large values because the frontal area involved along a displacement front is very small.

Water—Universal Contaminant

When water is considered as a factor in mode of failure, we are confronted with the fact that water is almost universally used as the reference liquid for determining if a solid surface is clean enough to be bonded, regardless of the nature of the adhesive to be used. This determination is made either by measuring the contact angle with water or by using the water break test after the cleaning cycles usually specified for the material to be bonded.

The final factor in defining a mode of bond failure based on water adsorption is that water may well be described as the universal contaminant; it is present everywhere in varying concentrations, and it has a finite permeability for most organic polymers.[44,45]

To summarize, adhesive failure of a properly prepared bond in which an adhesive is physically adsorbed on a polar surface may be accounted for as follows: Water, the universal contaminant, has a greater capability than most liquids for adhesion to most solids, especially polar solids. It can preferentially desorb most physically adsorbed liquids from most polar solids. Since fluid adhesives develop stresses at the interface as they solidify, the initial adhesion becomes less; therefore, the tendency for water to displace the adhesive becomes greater rather than less because of the stored interfacial stresses. Stress relief occurs by cleavage along the plane of the stress. This has been demonstrated by experiments with partly cleaved, stressed mica.[46,47] When the material is exposed to water vapor or immersed in water, the stress is relieved by further separation of the mica to a position involving a lower stress-level. This desorption by water and preferential adsorption of water can occur only at the solid/adherend surface where this preferential situation exists. Considerable displacement forces are developed in adsorption/desorption systems. Water is usually the reference liquid used to determine bondability of a high energy or polar surface. Thus, the surface condition is specifically tailored to make water the desorbing agent.

Relative Affinities

A weak boundary layer may account for adhesive failure if the liquid or volatile low-molecular-weight content of either the adhesive or adherend is incompatible and tends to bloom or bleed to the surface, or if the liquid content, being compatible and evenly-distributed throughout the adhesive, is preferentially adsorbed on the adherend surface, thereby desorbing the adhesive. If the adhesive has greater affinity for the adherend surface than the low-molecular-weight content of the adhesive or adherend, in terms of work of adhesion, then adhesion can be independent of the low-molecular-weight material sometimes designated as a "weak boundary layer."

On the basis of the values for work of adhesion in Table 2-2, the work of adhesion of water for a polar surface would be much greater than that of a resin system such as an epoxy used as an adhesive. Such resins are usually much more viscous and not as highly-oriented as the representative liquids evaluated in Table 2-2.

The use of the contact angle of water to determine cleanliness and bondability of a metal surface can be misleading and possibly harmful, especially if the bonded assembly is expected to withstand an aqueous environment in actual service. Once a metal surface has been certified as being clean by the water contact-angle method, the next step is to treat the surface with a chemically active adhesion promoter so that the surface is more receptive to the intended adhesive system than it is to water. This means that where it is possible to measure the contact angle of the adhesive, as in the case of unfilled epoxy resins, one should make sure that the adhesive has a lower contact-angle than does water for the adherend surface. Where adhesive contact-angles are not measurable, as in the case of thixotropic or film adhesives, laboratory experiments will be necessary to determine the optimum surface treatment.

SUMMATION

The discussion of the mechanism for adhesion failure in the presence of water may be summarized as follows:

1. On a metal-oxide surface, two forces are at work: (a) a force for desorption derived from the work of adhesion for water minus the work of adhesion for the adhesive, which should be a large positive figure (see Table 2-2); and (b) a force based on the release of stored interfacial stresses[46,47] in the presence of a wetting liquid.

2. These forces probably act along a linear front, since desorption proceeds faster at an adhesive/adherend interface than by permeation through the resin. Laird showed that diffusion of water along the glass/resin interface of an epoxy resin bonded to E-glass was 450 times faster than permeation through the resin.[48] Thus, the actual area of desorption along a front of molecular thickness would be very small, approaching zero.

3. The resultant stress available for desorption—that is, desorption forces divided by a very small desorption area—could be very large. Thus, given the proper combination of a physically adsorbed adhesive on a metal or metal oxide surface in an aqueous environment, adhesive failure is inevitable. Assuming that all solids are essentially very viscous liquids, all that is required is time to overcome the large viscosities involved.

   Initial testing of a properly prepared bond may measure cohesive strength whether in the adhesive or the adherend. However, after desorption has taken place because of environmental exposure, and when adhesive failure is evident, the measured strength is probably due to the attractive forces acting across the preferentially adsorbed water.

EXPERIMENTAL DATA

Reports on the durability of adhesives exposed to aqueous environments and elevated temperatures show that time to failure and type of failure vary depending on liquid or vapor aqueous environment, temperature, and stress condition.

Humidity and Temperature

Falconer et al. showed that structural bonds exposed to 80C (176F) and 100 percent relative humidity were approaching zero strength asymptotically after seventeen days.[49] In this case the initial failure of the bonds was obviously cohesive, while adhesive failure occurred after about seven days. The authors suggested that the equilibrium condition was being approached, i.e., all the available sites in the boundary layer being occupied by water.

Accelerated Stress

Studies of bonds prestressed during accelerated aging under conditions of high humidity and temperature show that the load at which the bonds failed was a fraction of the control test values for the unaged specimens.[50] It is possible that the initial failure was in adhesion because of preferential adsorption of water at the stressed adhesive/adherend interface. When the strength of the residual adhesive bond equalled the applied load, failure resulted at a fraction of the control value.

Time

A study of two structural adhesives bonded to aluminum showed that either adhesive or cohesive failure is possible after an initial cohesive failure.[51] The two adhesives selected for the study were (1) a nitrile rubber phenolic system known to be water resistant and (2) a nylon epoxy system known to be water sensitive. They were both film adhesives with liquid primers. After priming, preassembly drying, and assembly, they were cured for 90 minutes at 177C (350F), and 25-psi pressure. The aluminum used in the lap shear specimens was treated with a sulfuric acid dichromate etch per MIL-A-9067 before priming. One set of specimens of each kind was exposed to continuous humidity (> 93% RH) with the temperature cycling between 30C (86F) and 65C (149F) in a 48-hour cycle. Another set of specimens was immersed continuously in distilled water at ambient laboratory temperature [approximately 24C (75F)]. The specimens were not allowed to dry before testing. The results for the nitrile rubber phenolic are shown in Fig. 2-2, and for the nylon epoxy, the results are shown in Fig. 2-3.

The nitrile rubber phenolic specimens exhibited cohesive failure throughout two years of the test. The initial drop in strength was

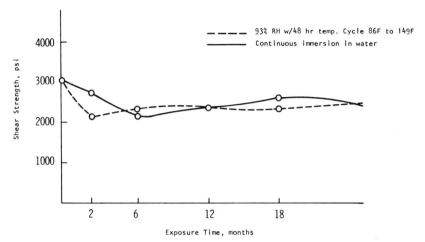

Fig. 2-2.   Effect of aqueous exposure on tensile shear strength of nitrile rubber phenolic adhesive.

probably due to the achievement of equilibrium with absorbed water. There was no evidence of adhesive failure, which indicated that the adhesive bond was due to hydrogen bonding or chemisorption.

On the other hand, nylon epoxy specimens went from initial cohesive to adhesive failure within the first two months. Specimens exposed to humidity leveled off in about two months to about 1100

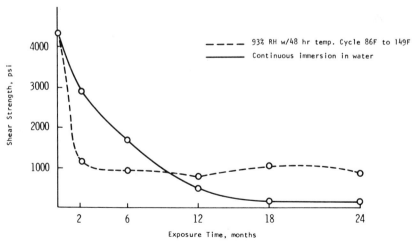

Fig. 2-3.   Effect of aqueous exposure on tensile shear strength of nylon epoxy adhesive.

psi and dropped only slightly for the remainder of the test; but, obviously, water had penetrated to the interface and desorbed the adhesive.

The immersed nylon epoxy specimens had a more gradual drop for about 18 months, with an asymptotic approach to zero for the last six months of the test. The final values averaged 120 psi.

These results, reported by Falconer,[49] Sharpe,[50] and DeLollis,[51] indicate that the type of failure and time to failure are influenced by adhesive/adherend systems, temperature, state of stress, and condition of exposure (that is, complete immersion or high humidity).

Corrective Measures

In view of all the factors influencing adhesion, including the fact that water is probably the major degradative agent, it would seem that the principal guide in adhesives development would be to use water only as an indication of surface cleanliness and not as the reference liquid for surface bondability. Surfaces should be modified by chemisorbed adhesion promoters which would have greater affinity for the intended adhesive system than for water. It would then be possible to design for greater durability. If the service environment included liquids or gases other than water or water vapor, then the development guidelines would be modified accordingly. Once the adsorption-desorption mechanism of bond failure is understood, the corrective measures become apparent.

The next chapter will describe three areas of corrective action in which surface treatments have been used to make adhesive bonds more durable.

REFERENCES

1. White, M. L. *The Detection and Control of Organic Contaminants on Surfaces.* New York: Marcel Dekker, Inc., to be published 1970.
2. Kline, G. M., and Reinhart, F. W. "Fundamentals of Adhesion." *Mechanical Engineering*, 72, 717. September 1950.
3. Bikerman, J. J. *The Science of Adhesive Joints.* New York: Academic Press, 1961.
4. Eley, D. D. *Adhesion.* London: Oxford University Press, 1961.
5. Sterman, S., and Toogood, J. B. "How to Promote Adhesion with Silicones and Silanes." *Adhesives Age*: July 1965.
6. Union Carbide Corporation. *Unlocking the Secrets of Silicones.* Product Information Bulletin. New York: Union Carbide Corp., August 1965.
7. Sterman, S., and Marsden, J. G. "Silane Coupling Agents as Integral Blends in Resin Filler Systems." *Modern Plastics*: July and October, 1963.

8. Bikerman, J. J. "Effect of Impurities on Polyethylene Adhesion." *Applied Chemistry* 11 (1961): 81–85.
9. Hansen, R. H., and Schonhorn, H. "A New Technique for Preparing Low Surface Energy Polymers for Adhesive Bonding." *Journal of Polymer Science*, March 1966: 203–209.
10. Sharpe, L. H., and Schonhorn, H. "Surface Energetics, Adhesion and Adhesive Joints." *Advances in Chemistry*, 43 (1964): 189–201.
11. Black, Judith A.; Lyman, D. J.; and Parkinson, D. B. *RTV Silicone Adhesives and Potting Compounds.* SRI Project PRD 5046, Special Report No. 2, July 1965.
12. Zisman, W. A. "Influence of Constitution on Adhesion." *Industrial and Engineering Chemistry* 55 (October 1963): 19–38.
13. Bikerman, J. J. "The Fundamentals of Tackiness and Adhesion." *Journal Colloid* 2 (1947): 174.
14. DeBruyne, N. A. *The Extent of Contact Between Glue and Adherend.* Bulletin No. 168. Cambridge, England: Aero Research Ltd., 1956.
15. Eley, D. D. "Surface Chemistry Studies in Relation to Adhesion." *Kolloid-Zeitschrift und Zeitschrift für Polymers*, Vol. 197, Book 1-2 (1964): 129–134.
16. DeBruyne, N. A. "The Aircraft Engineer," XVIII, No. 12, 53, December 28, 1939.
17. Moser, F. "Bonding Glass." *Handbook of Adhesives.* New York: Reinhold Publishing Corporation (1962): 523–33.
18. Zisman, W. A. "Constitutional Effects on Adhesion and Abhesion." NRL Report 5699, November 29, 1961.
19. Hodgman, Charles D., ed. *Handbook of Physics and Chemistry.* The Chemical Rubber Publishing Co., 1962.
20. DeLollis, N. J. "Structural Metal Bonding." *Handbook of Adhesives.* New York: Reinhold Publishing Corporation (1962): 513.
21. Schonhorn, H., and Sharpe, L. H. "Surface Energetics, Adhesion and Adhesive Joints II." *Polymer Letters* 2 (1964): 719–722.
22. Fowkes, F. M. "Surface Chemistry." *Treatise on Adhesion and Adhesives.* New York: Marcel Dekker, Inc. (1967): 325–449.
23. Laird, J. A., and Nelson, F. W. "The Effect of Glass Surface Chemistry on Glass Epoxy Systems," *SPE Trans.*, April 1964.
24. Erickson, P. W.; Volpe, A.; and Cooper, E. R. "Effects of Glass Surfaces on Laminating Resins." *Modern Plastics*, August 1964.
25. Outwater, J., and Kellogg, D. *A Simple Experiment to Show the Origin of Water Influencing the Debonding of Resin Against Glass*, Report for Contract No. V-3219 (01) (X), September 15, 1961.
26. Zisman, W. A. "Surface Chemistry of Glass-Fiber-Reinforced Plastics. NRL Report 6083. Presented at the Symposium on Glass-Resin Interface of the Society of Plastics Industry, Inc., 19th Annual Exhibit and Conference, Chicago, Illinois, February 6, 1964.
27. Schonhorn, H., and Hansen, R. H. "Surface Treatment of Polymers for Adhesive Bonding," *Journal of Applied Polymer Science*, 11 (1967): 1461–1474.
28. Schonhorn, H., and Ryan, F. W. "The Effect of Morphology in the Surface Region of Polymers on Adhesion and Adhesive Joint Strengths." ACS (American Chemical Society) Division of Organic Coatings and Plastics Chemistry, 154th Meeting, Vol. 27, No. 2, September 1967.
29. Henderson, A. W. "Pretreatment of Surfaces for Adhesive Bonding." In *Aspects of Adhesion*, edited by D. J. Alner. London: University of London Press, March 1963.
30. Panek, J. R. "Polysulfides, Sealants and Adhesives." *Handbook of Adhesives.* New York: Reinhold Publishing Corporation (1962): 286.

31. Brown, H. P., and Anderson, J. F. "Nitrile Rubber Adhesives." *Handbook of Adhesives*. New York: Reinhold Publishing Corporation (1962): 220.
32. Schollenberger, C. S. "Isocyanate Bond Adhesives." *Handbook of Adhesives*. New York: Reinhold Publishing Corporation (1962): 333.
33. O'Rear, J. J., and Sniegoski, P. J. "Chlorophenyl Alkyl-Substituted Carboxylic Acids and Silanes Designed as Adhesion Promoters." *ACS*: 4–10, September 1967.
34. Shafrin, E. G., and Zisman, W. A. "Preparation and Wettability of Terminally Chlorophenyl-Substituted Carboxylic Acid Films." *ACS*. September 1967.
35. Bascom, W. D. "The Wettability of Ethyl and Vinyl-Triethoxy-Silane Films Formed at Organic Liquid/Silica Interfaces." *ACS*: (September 1967), 27–42.
36. Young, F. M., and Rouch, L. L. "Surface Chemistry of Adhesion Promoters." *ACS*: (September 1967), 110–116.
37. Boyd, G. E., and Livingston, H. K. *Journal of the American Chemical Society*, 64 (1942): 2383.
38. Loeser, E. H.; Harkins, W. D.; and Twiss, S. B. *Journal of Physical Chemistry*, 57 (1953): 251.
39. Basford, P. R.; Harkins, W. D.; and Twiss, S. B. *Journal of Physical Chemistry*, 58 (1954): 307.
40. Harkins, W. D., and Dahlstrom, R. "Wetting of Pigments and Other Powders." *Industrial Chemistry*, 22 (1930): 897.
41. Smith, W. N. "Mathematical Analysis of Gel Permeation Chromatography Data." *Journal of Applied Polymer Science*, 11, (1967): 639–657.
42. Bartell, F. E., and Osterhof, H. J. "The Measurement of Adhesion Tension Solid Against Liquid." *Colloid Symposium Monograph*, V (1937): 113–134.
43. DeLollis, N. J. "Note on the Displacement Pressure Method for Measuring the Affinity of Liquids for Solids." *Journal of Physical Chemistry*, 56 (1952): 193.
44. Lee, H. "Attainment of Equilibrium in Water Absorption Tests of Epoxy Casting Resins." Paper read at the 4th Pacific Area American Society for Testing and Materials, National Meeting, 1962.
45. Peffer, R. J., and Dunbar, R. E. "Permeability and Absorption of Epoxy Films." *North Dakota Academy of Science Annual Proceedings*, 6 (1952).
46. Bowden, F. P. "The Adhesion of Metals and the Influence of Surface Contamination and Topography." In *Adhesion and Cohesion*, edited by Philip Weiss. New York: American Elsevier Publishing Company, Inc., 1962.
47. Bailey, A. I., and Kay, S. M. "A Direct Measurement of the Influence of Vapour, of Liquid and of Oriented Monolayers on the Interfacial Energy of Mica." *Proceedings of the Royal Society*. A301 (1967): 47–56.
48. Laird, J. A. *Glass Surface Chemistry for Glass Fiber Reinforced Plastics*, Final Report, Navy Contract No. W-0679-C (FBM), June 1963.
49. Falconer, D. J.; MacDonald, N. C.; and Walker, P. "The Effect of High Humidity Environment on the Strength of Adhesive Joints." *Chemistry and Industry*, July 1964: 1230.
50. Sharpe, L. H. "Prestressed Bonds, Aspects of the Permanence of Adhesive Joints." Paper read at Symposium on Structural Adhesives Bonding, September 14–16, 1965, at Stevens Institute of Technology Center, Hoboken, New Jersey.
51. DeLollis, N. J. "Mode of Failure in Structural Adhesive Bonds." *Journal of Applied Polymer Science*, June 1967.

CHAPTER 3

Cleanliness and Surface Treatment of Metals[*]

THE CONCEPT OF CLEANLINESS

An old saying has it that cleanliness is next to godliness, and this is as applicable to bondable surfaces as it is to people—whether we're concerned with metal surfaces or surfaces of any other material in the environmental shell surrounding the earth. High-vacuum equipment at elevated temperatures is often utilized in basic research on clean surfaces. The use of really exotic equipment probably is best exemplified in the work of Dr. Muller with his field ion microscope.[1]

Equilibrium Cleanliness

At the practical level of industrial bonding, a clean surface may be defined as one which, after being degreased, etched, and rinsed in distilled or deionized water, is dried with clean air. The surface then comes to equilibrium with its environment before bonding, and is considered "clean."

Equilibrium Environment

The cleanliness of the equilibrium environment is dependent on the criticality of the bonding operation. At its worst, the equilibrium environment may consist of air filled with not only the smog, oil, grease, and acid of an industrial area, but also with laboratory contaminants such as vacuum-pump exhausts, acid vapors, and aerosol-propelled vapors of silicone, fluorocarbon, or wax-type release agents. White[2] has shown that the time for a metal surface to come to equilibrium with airborne contaminants is about five hours (see Fig. 2-1).

---

[*] Portions of this chapter were published in *Adhesives Age*, December 1968-January 1969.

The basic approach to improving the environment is to isolate the cleaning and bonding areas from other operations, preferably in separate rooms. If separate rooms are not available, individual compartments in a corner of a room or on an assembly line will suffice, as long as a source of dry, filtered air provides a positive flow, away from the cleaning and bonding operations.

The cleaning and bonding operations should be isolated from each other. The metal-cleaning area should contain only the tanks, acids, solvents, and rinsing baths necessary to the operation. The bonding areas should contain only those materials needed to apply the adhesive and assemble the parts, including the tooling necessary to hold the surfaces together during cure.

A prime requirement for all these operations is that the processing personnel develop a concept of cleanliness. Process specifications are ineffective unless the operator works with due care and understanding.

CLEANLINESS VARIABLES

A solvent from an unopened can, fresh from the storeroom, is contaminated if it is poured into a dirty container. A wiping cloth dipped repeatedly into clean solvent may leave more dirt on the metal than it removes. (Solvent should be poured on the cloth, so that the cloth never comes in contact with the solvent in the container.)

Degreasers

Items cleaned in degreasers should be exposed to the vapors so that there is a maximum of condensation and washing action.[3] Part surfaces should not be in contact. Parts should not be soaked in the degreaser sump, since this not only collects dirt on the surface but heats up the part so that condensation is minimized.

Sandblasting

Sandblasting and vapor-honing operations should be done after degreasing and before ultrasonic cleaning for the following reasons:

1. Prior degreasing means that contamination of the sandblasting and vapor-honing facilities will be minimized.
2. Sandblasting and vapor honing will probably leave particulate matter on the surfaces, and this can subsequently be removed by the ultrasonic cleaning.

Since the sandblasting and vapor-honing operations may release airborne particles, they should be isolated from other cleaning operations. Compressed air or inert gas used for blowing off particulate matter after sandblasting or for drying surfaces should be dry and oil-free.

It should be noted that "ultrasonic cleaning" becomes a misleading magic phrase if the cleaning solvents are not renewed frequently.

Etching Baths

Acid or alkaline etching solutions should be monitored regularly to prevent excessive build-up of contaminant by-products and to maintain the proper chemical concentration. Tank temperatures should be kept within specification requirements so that reaction rates are constant. Continuous and vigorous circulation of the solutions is necessary for uniform treatment. A scummy surface is a danger signal which requires immediate attention.

Surface Protection and Storage

Once cleaned, items may have to be stored for hours, or even for days before being bonded. Because of this, items should be individually packaged in dustproof packages in clean areas. If possible, especially for sensitive surfaces such as steel, copper, or iron alloy, it would be desirable to coat the surfaces to be bonded, immediately after cleaning, with a primer compatible with the intended adhesive system. Quite a few adhesives are supplied with a recommended primer, so that for stored items the primer would serve a dual function: it would protect the metal surface, and at the same time it would probably be more easily wetted by the adhesive during the bonding process than would the original metal surface.

Chessin and Curran note typical storage-life for various metals subjected to different treatments, as follows: [4]

Aluminum	wet-abrasive blasted	72 hours
Aluminum	sulfuric, chromic-acid etch	6 days
Aluminum	anodized	30 days
Stainless steel	sulfuric-acid etched	30 days
Steel	sandblasted	4 hours
Brass	wet-abrasive blasted	8 hours

The short storage-life of steel and brass emphasizes the sensitivity of those surfaces.

GENERAL CLEANING METHODS

Solvent Wiping

A widely used cleaning method—but probably the least satisfactory—is wiping with clean solvent applied with clean, lint-free cloth or paper. The concept of cleanliness becomes very important here, in that the operator must be continually on the alert for sources of contamination. The solvent should be poured or squirted onto the wiping cloth; the cloth should never be allowed to touch the mouth of the container. The part should be wiped until no trace of dirt is noticed on the wiping cloth; streaks remaining on the wiped surface indicate that the cloth or the solvent was not clean to begin with.

Acceptable solvents for wiping are toluene, methyl-ethyl ketone, trichloroethylene, and acetone.

For small parts, rinsing in several changes of clean solvent and blowing off the parts in clean dry nitrogen would be better than wiping.

Vapor Degreasing

Vapor degreasing is preferable to wiping, since here the surfaces to be bonded are continually washed with the distilled, condensed vapors of the solvent, usually trichloroethylene or perchloroethylene, at the boiling temperature of the solvent. Surface soils are concentrated in the sump, so parts should not be dipped into the sump.

The vapor temperature should be monitored since some contaminants may increase the boil temperature of the sump solution, thereby volatilizing some high-boiling contaminants which may eventually form an objectionable film on the bonding surface.

Catalyzed resin on a metal part should first be rinsed in a solvent at ambient temperature, since the vapor temperature in a vapor degreaser might cure the resin in place before washing it away.

Abrasive Cleaning

Abrasive blasting with sand or grit is often used as a convenient positive method for removing tarnish, rust, or mill scale, especially for large parts. A grease or oil film on the parts should be removed by vapor degreasing or solvent wiping before the sandblasting operation to minimize contaminant carryover. The air used for sandblasting should be dry and filtered to eliminate water and oil. This is especially important in humid areas. The sand or other abrasive material should be clean, and should not be reused. Air pressure and type of grit should be varied depending on the hardness of the surface being cleaned.

Solid particles left on the surface after abrasive blasting can be removed with clean, dry, compressed air. Solvent wiping can also be used, but in view of the aforementioned uncertainties inherent in this method, it should be used only as a last resort or if there is considerable handling of the blasted surface before the bonding operation. Priming of the freshly sandblasted surface should also be considered.

Mechanical Working of Adhesive

Another possibility, not often considered, is that if a primer or adhesive is to be brushed on, it should be applied with a stiff-bristle brush using as much mechanical energy as possible. This helps to work the primer or adhesive into the surface so that particle contaminants and even surface films are suspended in the adhesive or primer. The adhesive or primer will then wet the surface more thoroughly.

Some work has been reported in which an attempt was made to correlate strength values of bonded specimens with the mechanical energy expended in rubbing the adhesive into the metal surface.[5] The results indicate that when the adhesive was rubbed in, the specimens exhibited higher strength values with less scatter. One might conclude that rubbing in the adhesive removes some of the surface contamination and allows for better wetting. The disadvantage of this technique is that it would be too inconvenient and time-consuming for most production schedules. It might merit consideration in special cases where conventional cleaning is impractical.

Abrasive blasting sometimes has definite advantages over other cleaning methods. Where the surface to be bonded comprises two or more parts of similar or dissimilar metals held together by mechanical fasteners, such as screws or rivets, for example, any solvents, acids, or alkaline material used for cleaning would fill the joints and be difficult to rinse out thoroughly. Corrosion could ensue, especially in the case of acids or alkaline solutions. Dry sandblasting followed by a dry-air blast to remove particulate contaminants would leave no such troublesome residue.

Vapor Honing

For many small or thin-gauge metal parts the usual abrasive blasting treatment would be much too rough. A milder, but still thorough abrasive cleaning method would be vapor honing, also known as vapor blasting or hydroblasting. The procedure consists of cleaning with finely-divided abrasive of varying mesh-size suspended in water, with a detergent and a corrosion inhibitor as additives. This method is good for cleaning small parts but does require thorough rinsing after the honing.

Ultrasonic Cleaning

The final step, to follow the cleaning methods already described, could very well be ultrasonic cleaning, in which a high-frequency generator coupled with a transducer produces sonic waves which are carried to the part being cleaned by the cleaning liquid in which the part is immersed. Details of design and construction are found in Reference 3 and in the industrial literature.

Ultrasonic cleaning is intended primarily for small parts. The parts to be cleaned must be suspended in the liquid and must not touch the tank bottom, since this deadens or distorts the sonic waves. Overcrowding of parts must be avoided, and the liquids should be changed frequently to prevent the formation of a contaminant film on the cleaned parts.

Various solvents and processes can be used in sequence with ultrasonic cleaners, depending on the type of soil being removed— the consideration being whether the soil is wet best by water, detergent in water, or some organic solvent such as trichloroethylene. Representative sequences are:

1. Detergent solution, water, alcohol, and vacuum bake
2. Trichloroethylene, alcohol, and vacuum bake.

Combinations of Cleaning Methods

Table 3-1 gives some idea of the efficacy of various cleaning methods, used singly and in various combinations, in terms of the lap shear strength of bonded aluminum specimens. The adhesive, a constant in every case, is a filled two-part epoxy. The results shown in Table 3-1 (taken from Reference 4) indicate that the best method —at least for the materials used in that particular survey—is the one that uses all procedures, including alkaline and acid etch, in the proper sequence. It is possible that the grit blasting increases the surface area while the etching conditions the surface. It is significant that, in this series, the large drop in strength values takes place at cleaning method 7 after the sulfuric acid dichromate method has been eliminated.

Table 3-1. Surface Preparation versus Lap-shear Strength[a,b]

Group Treatment	$\overline{X}_c$, psi	S, psi	C_v, percent
1. Vapor degrease, grit blast 90-mesh grit, alkaline clean, $Na_2Cr_2O_7$—H_2SO_4, distilled water	3091	105	3.5
2. Vapor degrease, grit blast 90-mesh grit, alkaline clean, $Na_2Cr_2O_7$—H_2SO_4, tap water	2929	215	7.3
3. Vapor degrease, alkaline clean, $Na_2Cr_2O_7$—H_2SO_4, distilled water	2800	307	10.96
4. Vapor degrease, alkaline clean, $Na_2Cr_2O_7$—H_2SO_4, tap water	2826	115	4.1
5. Vapor degrease, alkaline clean, chromic-H_2SO_4, deionized water	2874	163	5.6
6. Vapor degrease, $Na_2Cr_2O_7$—H_2SO_4, tap water	2756	363	1.3
7. Unsealed anodized	1935	209	10.8
8. Vapor degrease, grit blast 90-mesh grit	1751	138	7.9
9. Vapor degrease, wet and dry sand, 100- + 240-mesh grit N_2 blown	1758	160	9.1
10. Vapor degrease, wet and dry sand, wipe off with sandpaper	1726	60	3.4
11. Solvent wipe, wet and dry sand, wipe off with sandpaper (done rapidly)	1540	68	4
12. Solvent wipe, sand (not wet and dry), 120 grit	1329	135	1.0
13. Solvent wipe, wet and dry sand, 240 grit only	1345	205	15.2
14. Vapor degrease, aluminum wool	1478		
15. Vapor degrease, 15 percent NaOH	1671		
16. Vapor degrease	837	72	8.5
17. Solvent wipe (benzene)	353		
18. As received	444	232	52.2

[a] Resin employed is Epon 934 (Shell Chemical Company); cured 16 hours at 24C (75F) plus 1 hour at 82C (180F). Fillet on overlap left intact, adhesive on sides of specimens removed.

[b] Author[4] indicates that 10 to 20 test specimens were used to establish the value for each of the cleaning conditions.

[c] $\overline{X}$ = average value; S = standard deviation; C_v (percent) = $\dfrac{S}{\overline{X}} \times 100$ = coefficient of variation.

Rinse Water

A recent paper by Wegman et al. describes the effect of rinse water on the bonded joint.[6] It shows that final rinsing of an acid-etched aluminum surface in deionized water results in significantly-decreased bond strengths as compared with rinsing in water containing divalent ions. Thus it cannot be assumed that deionized or distilled water is the best for final rinsing.

Up to this point, I have not discussed alkaline and acid etching since these tend to be specific for the metal surface being prepared. They will be discussed in the section entitled "Specific Cleaning Processes."

COMPROMISES IN THE CLEANING PROCESS

A final point to note is that, in the solution of bonding problems as in other problem areas, we are often faced with the need to compromise.

An outstanding example of compromise is in the automobile industry, where for many years the hood stiffener has been bonded to the hood with a vinyl resin adhesive. At the time of bonding, the stamped metal parts are still coated with oil, which is necessary to prevent the very sensitive clean steel surface from rusting.[7] The adhesive is formulated to tolerate the oil—possibly by absorption—during the bonding process. Once assembled, the hood is cleaned and painted. The cure of the adhesive takes place when the paint is baked. The bond is expected to last for the life of the car.

Other compromises must be expected: sometimes extended time must be allowed between cleaning and bonding; perhaps you must choose from a variety of cleaning procedures, such as those listed in Table 3-1; or a cleaning process may have to be fitted into a production schedule.

Bond Contaminants

Table 3-2 gives an indication of the amount of contamination that a bond can tolerate, but this may be very misleading, since no data were accumulated concerning the durability of the various bonds when exposed to degrading environments. It is interesting that low strength values and adhesive failures occurred most frequently as a result of (1) using a silicone oil contaminant, (2) exposure to high humidity, and (3) wiping with a solvent.

Economics

In the final analysis, the choice of a compromise will be based on the economics of the situation. A process need only be good enough for the job at hand, but sufficient exploratory and development research should be done to insure that it is satisfactory before the final choice is made.

SPECIFIC CLEANING PROCESSES

Aluminum

Aluminum is probably the metal most widely used in structural bonding. Its strength, low density, corrosion resistance, and abundance have given it a virtual monopoly in the aircraft industry. Even if it cannot be used for supersonic aircraft, because of increased skin temperature due to atmospheric friction, aluminum will still be useful in Mach 2 aircraft. Bonded aluminum is also commonly used in building panels, light poles, skis, seating equipment, refrigerator parts, etc.

The general cleaning methods previously described are all applicable, depending on the type of bond expected (as shown in Table 3-1). The methods used specifically for aluminum are:

I. Sulfuric acid dichromate etch used by the aircraft industry and specified in MIL-A-9067:
   a. Solvent clean in perchloroethylene or trichloroethylene (vapor degreasing is preferable if available)
   b. Air dry to keep solvent from contaminating acid bath
   c. Submerge aluminum in acid bath for 10 minutes at bath temperature of 66–68C (150–155F)
      Chemical concentrations in bath:
      Concentrated sulfuric acid–10 parts by wt
      Sodium dichromate–1 part by wt
      Distilled or deionized water–30 parts by wt
   d. Rinse off acid in fresh running water, and air-dry or bake-dry at temperatures up to 66C (150F).

   NOTE: A measure of the success of the etching and cleanliness of the rinse water is a clean water break with no indication of water running off in rivulets or gathering into drops.

Continuous check should be made of acid-bath temperature and chromate ion concentration.

Table 3-2. Tensile Adhesion Strength of Bonds[a] with Varying Degrees of Contamination

Specimen Number[b]	Surface Condition	Glue-Line Thickness (inches)	Bond Strength, psi	Type of Failure
1. 1–8	Machined surface dipped in sulfuric acid-sodium dichromate solution at 66 to 77C (150 to 170F) for 5 to 10 minutes, rinsed allowed to dry, then bonded	0.005	Avg of 8 specimens 5042	Cohesion
2. 9–10	Same as above, plus SAE-30 oil on surface	0.005	Avg of 2 specimens 4487	Cohesion
11–12	Same as No. 1, plus exposure to 90% humidity for 24 hrs	0.005	Avg of 2 specimens 1577	Adhesion
13–14	Same as No. 1, plus light film of DC-200 silicone oil	0.005	Avg of 2 specimens 1077	Adhesion
15–16	Same as No. 1, plus light film of cutting fluid	0.005	Avg of 2 specimens 5025	Cohesion
3. 17–24	Sandblasted surface, sand blown off	0.005	Avg of 8 specimens 4307	Cohesion
4. 25–26	Sandblasted, plus SAE-30 oil	0.005	Avg of 2 specimens 5735	Cohesion
27–28	Sandblasted, plus exposure for 16 hrs at 90% humidity	0.005	Avg of 2 specimens 2527 psi	Adhesion

Table 3-2 (Continued).
Tensile Adhesion Strength of Bonds[a] with Varying Degrees of Contamination

Specimen Number[b]	Surface Condition	Glue-Line Thickness (inches)	Bond Strength, psi	Type of Failure
29–30	Sandblasted, plus light film of DC-200 grease	0.005	Avg of 2 specimens 454	Adhesion
31–32	Sandblasted, plus light film of cutting fluid	0.005	Avg of 2 specimens 5230	Cohesion
5. 33	Same as No. 3	0.001	4630	Cohesion
35	Same as No. 3	0.005	4700	Cohesion
36	Same as No. 3	0.008	5020	Cohesion
37	Same as No. 3	0.010	4545	Cohesion
38	Same as No. 3	0.015	4905	Cohesion
39	Same as No. 3	0.020	5240	Cohesion
40	Same as No. 3	0.030	4550	Cohesion
41	Same as No. 3	0.040	4470	Cohesion
6. 42–43	Sandblasted surface; 1/4" holes drilled on bonded surface	0.005	Avg of 2 specimens 4888	Cohesion
7. 44	Anodized aluminum to anodized aluminum, wiped clean with acetone	0.005	1345	Adhesion
8. 45–46	Anodized aluminum to machined aluminum, wiped clean with acetone		Avg of 2 specimens 3148	Adhesion to anodized surface

[a] Bonds were prepared from aluminum plugs using a filled-epoxy adhesive per MIL-A-8623, Type II cured with diethylaminopropylamine at 66C (150F) for 6 hours.
[b] All specimens were cured at 66C (150F) for 6 hours.

Tanks should be lead lined and equipped with an exhaust system to lead off solvent and acid vapors.

II. A chromic-acid method similar to Method I:
   a. As in I-a
   b. As in I-b
   c. Submerge aluminum for 5 minutes in a 71–82C (160–180F) solution of the following concentrations:
      Concentrated sulfuric acid–1 gallon
      Chromic acid–45 ounces
      Distilled or deionized water–9 gallons
   d. As in I-d.

III. A chromate method using a cold solution:[8]
   a. Degrease aluminum with a 50/50 solvent solution of methyl-ethyl ketone and chlorothene
   b. Abrade lightly with mildly abrasive cleaner
   c. Rinse in deionized water; wipe, or air-dry
   d. Etch 20 minutes in following solution:
      Sodium dichromate–2 parts by wt
      Sulfuric acid, 96%–7 parts by wt
   e. Rinse thoroughly in a deionized water spray; dry in circulating-air oven at 70C (158F) for 30 minutes.

Picatinny Arsenal[9] lists six, aluminum-cleaning methods used with an epoxy novolac adhesive, with the results shown in Table 3-3. Each value is the average of five specimens. It should be noted that tensile adhesion values are about twice as high as lap-shear values.

The ASTM recommended practice for the preparation of metal surfaces for adhesive bonding lists other methods in addition to those described herein. It also recommends a paste composed of solutions per Methods I and II, mixed with finely-divided silica or fuller's earth to be applied to aluminum surfaces which do not lend themselves to dipping. With a paste application, extra care must be taken in wiping off the treated surface with water-dampened cloths.

Copper and Copper Alloys

Prebonding treatments for copper and for copper alloys are usually the same, since the common denominator is the very sensitive nature of the copper. While brass is less sensitive to corrosive conditions than copper, all cleaning methods emphasize the importance

Table 3-3. Results of Six Aluminum-cleaning Methods

Chemicals	Concentration, gms	Application and Temp.	Tensile Adhesion, psi	Variance
A. HCl 37%	200	Cold dip	3090	33.2
Distilled water	540	. . .	. . .	. . .
Potassium dichromate	8	. . .	. . .	. . .
B. Per MIL-A-9067	. . .	68C (155F)	5790	7.3
C. HCl 37%	500	Cold brush-on	6500	4.9
Distilled water	200	. . .	. . .	. . .
Sodium dichromate	20	. . .	. . .	. . .
D. NaOH	90	Cold dip	. . .	. . .
Distilled water	510	. . .	. . .	. . .
followed by				
Nitric acid (HNO^3)	30 cc	. . .	. . .	. . .
Hydrofluoric acid	3	Cold dip	6730	7.9
Distilled water	567	. . .	. . .	. . .
E. Nitric acid	30 cc	. . .	. . .	. . .
Hydrofluoric acid	3	Hot dip	5500	4.6
Distilled water	567	. . .	. . .	. . .
F. Sandblast	. . .	. . .	6540	5.6

of applying the adhesive and bonding as soon as possible after the cleaning operation.

Since some adhesives have a tendency to be corrosive, it is good practice, if a durable humidity-resistance bond is expected, to run tests beforehand to check the corrosive effect of the adhesive on the copper or brass adherend. For instance, epoxy resins cured with triethylene tetramine or diethylene triamine are corrosive to copper and brass, especially in a humid environment.

ASTM recommended practice D-2651 lists five methods for treating copper and copper alloys. Three representative methods are repeated here:

  I. a. Degrease—again, vapor degreasing is preferred to wiping
     b. Immerse for 1 to 2 minutes at room temperature, in the following solution:
        Water–197 parts by wt
        Nitric acid (sp gr 1.5)–30 parts by wt
        Ferric chloride solution (42%)–15 parts by wt

    c. Rinse thoroughly, and dry as soon as possible with clean, dry air or nitrogen at room temperature to minimize tarnishing

    d. Apply adhesive immediately.

II. a. As in I-a

    b. Bright-dip at room temperature in concentrated nitric acid for 15 seconds or until all corrosion has disappeared

    c. As in I

    d. As in I.

III. This method is for copper alloys containing over 95 percent copper. It tends to leave a more stable surface especially if heat is used in the bonding operation as when used with molten polyethylene.

    a. Degrease

    b. Immerse for 30 seconds at room temperature in the following solution:

        Nitric acid (70% technical)–10 parts by vol

        Water–90 parts by vol

    c. Rinse in running water and transfer immediately to the next solution. Do not allow to dry.

    d. Immerse for 1 to 2 minutes at 98C (208.4F) in the following solution:

        Ebanol C (Enthone Inc., New Haven, Conn.), or equivalent—24 ounces to make one gallon of solution in water.

    NOTE: Do not boil the above solution.

    e. Rinse in running cold tap-water and air-dry. Bond as soon as possible within the working day.

This treatment leaves a black, surface finish which is nonconductive: i.e., the finish must be removed where an electrically conductive, solderable joint is needed.

ASTM recommended practice D-2651 includes a ferric sulfate/sulfuric acid treatment, a sulfuric acid/sodium-dichromate treatment, and an alkaline sodium chlorite/trisodium phosphate/sodium-hydroxide treatment.

Another source describes ammonium persulfate and ferric-chloride methods as follows: [10]

IV. a. Vapor degrease
   b. Immerse for 1 to 3 minutes at room temperature in the following solution:
      Ammonium persulfate–25 parts by wt
      Water–75 parts by wt
   c. Rinse in cold running water, dry in clean dry air, and apply adhesive as soon as possible.

V. a. Degrease
   b. Immerse 1 to 2 minutes at room temperature in the following solution:
      Hydrochloric acid (concentrated)–50 parts by wt
      Ferric chloride–20 parts by wt
      Water–30 parts by wt
   c. As in IV-c.

Good results with copper and brass have also been achieved with sandblasting or by using abrasive papers. Here again, the importance of applying the adhesive as soon as possible cannot be overemphasized. Bond strengths to be expected using an abrasive cleaning method are shown in Table 3-4.

Beryllium

Beryllium can be considered as a rather exotic addition to the family of structural metals. It is difficult to work and is considered toxic, thus, cleaning operations should be carried out in solution or with paste abrasives in hoods, eliminating the danger of inhaling air-

Table 3-4. Strength[a] of Conductive-Epoxy Adhesive[b] Using Brass Tensile-adhesion Plugs

Cure Time at 104C (220F), hours	Tensile-adhesion at Room Temperature, psi	Tensile-adhesion[c] at 127C (260F), psi
2	5000	1960
4	5800	2050
8	4000	2020

[a] Specimens consisted of 1⅛-inch diameter, 1-inch-long brass plugs bonded in pairs at 1⅛-inch diameter face. Surfaces prepared by abrading with crocus cloth and wiping with clean methyl-ethyl ketone immediately before applying adhesive.
[b] Adhesive consisted of unmodified-epoxy resin with 65 percent silver-flake filler. Curing agent was aromatic-amine eutectic with a concentration of 20 parts per 100 of epoxy resin.
[c] Failure was 100 percent cohesive in adhesive layer.

borne particles. Work areas should be covered with paper so that toxic remnants can be wrapped up and safely disposed of. After working with beryllium, one should carefully wash the face and hands.

In spite of its drawbacks, beryllium cannot be ignored, since its favorable strength-to-weight ratio makes it very useful for aerospace applications. As its production increases and the price drops, its use is bound to increase.

A recommended cleaning process for beryllium is as follows:

Beryllium cleaning process etch systems used:

  a. Etch solution:
    Sodium dichromate–66 parts by wt ($Na_2Cr_2O_72H_2O$)
    Sulfuric acid, 96%–660 parts by wt (sp gr 1.84)
    Water–1000 parts by wt
    Etch 30 to 60 sec in solution at 50 to 60C (122 to 140F).
  b. Etch solution:
    Ortho-phosphoric acid, 450 cc
    Concentrated sulfuric acid, 26.5 cc (H_2SO_4)
    Chromic anhydride 56.25 gm (CrO_3)
    Etch for 60 sec in solution at 110 to 120C (230 to 248F).

Procedure

  1. Degrease with "Lubtone" and water; distilled-water rinse
  2. Solvent degrease; reagent trichloroethylene
  3. Etch solution No. a
  4. Distilled-water rinse until pH paper shows neutral
  5. Etch solution b
  6. Repeat step 4
  7. Air-dry.

Beryllium, tensile lap-shear specimens cleaned by this method and bonded with an epoxy adhesive per MIL-A-8623, Type II, gave strength values averaging between 3000 and 4000 psi.

Other cleaning methods reported include vapor-blast resulting in 4500 psi bonds and use of a proprietary cleaning agent "Prebond 700,"[11] which gave similar results, all with the aforementioned type of adhesive.

Carbon-Steel Metals

Adhesive bonding finds wide acceptance as a fastening method for steel. Like copper and copper alloys, a fresh steel surface rusts very easily and must be protected by a primer compatible with the intended adhesive system, or else the adhesive must be applied to the freshly-cleaned surface as soon as possible. Once bonded, the exposed steel/adhesive joint should be coated with a corrosion-resistant coating to prevent corrosive attack.

The most common method for preparing steel surfaces for bonding is probably sandblasting, preceded by vapor degreasing to get rid of any oily film.

Since a clean steel surface is so easily oxidized, it is doubly important that the source of compressed air be dry and oil-free. Some methods call for rinsing in a clean, volatile solvent such as methylethyl ketone or isopropyl alcohol. These solvents usually contain water and should be used with caution, since their water content can cause steel to rust. When volatile solvents evaporate from a metal surface, they may chill the surface; in a humid environment, this may result in condensation of water, which nullifies the cleaning effort. The best treatment after sandblasting is blowing the particulate matter off the surface with clean, dry air. Solvent wiping should be used only as a last resort.

Vapor honing or vapor blast can be used for small steel parts without danger of warping or of any significant change in tolerance. The abrasive-water suspension usually includes a wetting agent and a rust inhibitor. These must be rinsed off thoroughly in clean water. The use of a clean, dry, water-compatible solvent is necessary here, to remove the water. Clean, dry, compressed air or nitrogen should be used to blow off the solvent.

Thin sheet-metal stock may warp when cleaned by abrasive methods. Any of the acid-etch solutions used with stainless steel can be used with carbon steels. They should be used with caution, however, since they would react more rapidly with carbon steel than with stainless steel.

Stainless Steel

Additional information on surface preparation of stainless steel is available in References 12 and 13. All cleaning should be preceded by vapor degreasing to get rid of oils or greases and other contaminating organic films. Abrasive treatment, such as grit or sandblasting or wire brushing, is very effective. After blasting, the surfaces should be blown off with clean, dry air or nitrogen. Vapor blasting with water or steam containing suspended abrasive particles is also very effective, especially with small items. Vapor blasting should be followed by rinsing in clean water and alcohol, and blowing off with clean, dry air or nitrogen. Sheet metal should be cleaned in etching solutions to minimize warping.

Uniform abrasive cleaning preceding etching operations seems to increase bond strength, possibly because of the increased surface area and the attendant bonding sites made available to the adhesive. Some work has been done to demonstrate the beneficial effect of roughness when a polyether urethane adhesive is used on nickel surfaces[14] (see Table 3-5).

Table 3-5. Effect of Roughness of Nickel Surface on Peel Strength

Roughness, microinches	Peel Strength, grams*	Roughness, microinches	Peel Strength, grams
1.7	90	4.5	135
2.0	90	6.0	180
2.5	105	9.0	250

* Bond width equals 5/8-inch.

Stainless-steel surfaces are not nearly as sensitive as those of carbon steel, so that some delay between cleaning and bonding may be tolerated; but the surfaces must be protected during the interim.

Representative cleaning methods are:

I. a. Degrease (vapor degreasing preferred)
   b. Etch for 10 minutes at 65 to 71C (150 to 160F) in the following solution:
      Water–90 parts by wt
      Sulfuric acid (sp gr 1.84)–37 parts by wt
      Nacconal NR or equiv.–0.2 part by wt
      Rinse thoroughly and remove smut with a stiff brush if necessary.
   c. Immerse in following bright-dip solution for 10 minutes at room temperature:
      Water–88 parts by wt
      Nitric acid (sp gr 1.5)–15 parts by wt
      Hydrofluoric acid (35.5%, sp gr 1.15)–2 parts by wt
   d. Rinse thoroughly in distilled or deionized water. Dry in oven at not over 93C (200F).

II. a. Degrease
   b. Immerse for 2 minutes at approximately 93C (200F) in the following solution heated by a boiling-water bath:
      Hydrochloric acid (sp gr 1.2)–200 parts by wt
      Phosphoric acid (ortho, sp gr 1.8)–3 parts by wt
      Hydrofluoric acid (35.5%, sp gr 1.15)–10 parts by wt
   c. Rinse thoroughly in distilled or deionized water. Dry in oven at not over 93C (200F).

III. a. Degrease
   b. Immerse for 15 minutes at 63 ± 3C (145 ± 5F) in the following solution:
      Sulfuric acid (sp gr 1.84)–100 parts by vol
      Saturated sodium-dichromate solution–30 parts by vol.
      (Approximately 75 parts by wt of sodium dichromate in 30 parts by wt of water).

NOTE: When preparing solutions of concentrated acid in water, always add the acid slowly and carefully to the water. Avoid splashing; wear safety-goggles and rubber apron.

   c. Rinse thoroughly in distilled or deionized water. Dry in oven at not over 93C (200F). Details on other cleaning methods are given in the aforementioned references.

Titanium

Another space-age metal which is achieving prominence in supersonic transport is titanium. This metal was not given much consideration until skin temperature for supersonic planes rose above 300F. Then it became the leading candidate for use when the best strength-to-weight ratio at elevated temperatures was required.

Since titanium is intended for application temperatures up to 600F, the surface preparations are directed toward bond-durability at temperatures up to 316C (600F). Treatments which have received the widest acceptance are based on nitric acid-hydrofluoric acid combinations. One of these is:[15]

I. a. Methyl-ethyl ketone wipe
   b. Trichloroethylene vapor degrease
   c. Pickle in the following water-solution:
      Nitric acid–15 percent by vol –70 percent HNO_3 sol
      Hydrofluoric acid–3 percent by vol of 50 percent HF sol

    d. Rinse in water at room temperature

    e. Immerse for 2 minutes in the following water-solution at room temperature:

        Trisodium phosphate–50 grams/liter of solution

        Potassium fluoride–20 grams/liter of solution

        Hydrofluoric acid (50 percent solution)–26 milliliters/ liter of solution

    f. Rinse in tap water at room temperature

    g. Soak in 66C (150F) tap water for 15 minutes

    h. Spray with distilled water and air dry.

II. Same as I, except for an additional final cleaning in alkaline detergent-sodium silicate solution.

Reference 15 lists 27 treatments with test results at temperatures up to 316C (600F). The two treatments described above, when evaluated with a nitrile phenolic and with an epoxy phenolic, give ambient results as follows:

Treatment	Adhesive	Tensile Lap Shear (psi)
I	Epoxy phenolic	2840
	Nitrile phenolic	2830
II	Epoxy phenolic	3380
	Nitrile phenolic	3840

Considerably more work has been done recently to improve the performance of titanium-bonded assemblies at temperatures up to 600F. The industries most knowledgeable in this field are the aircraft companies concerned with supersonic flight.

Present work in the field of titanium metallurgy has indicated that titanium has the property of absorbing considerable oxygen, which in itself creates a problem: The absorbed oxygen and the oxygen permeating from the unbonded surfaces attack the adhesive at the adhesive/adherend interface. Therefore, any metal treatment to be effective at elevated temperatures should seal the metal micropores to create an impermeable barrier to isolate the adhesive from the oxygen.

Magnesium

Magnesium, the lightest metal used in structural bonding, is also the most sensitive to corrosion; therefore, surface preparation of magnesium is intended primarily to prevent corrosion. Since the corrosion preventive coatings are cohesively weak, the major problem is to have them thick enough to prevent corrosion, but not so thick as to encourage failure within the coatings themselves.

Proprietary treatments for magnesium surfaces have been developed by magnesium-alloy producers such as the Dow Chemical Company. These treatments and surface dichromate conversion-coatings and wash-primers make magnesium surfaces suitable for bonding; but they should be evaluated to be sure they fit the intended application. Details are available from the magnesium producers, from the ASM Handbook, Volume II, and from the Military Specifications MIL-M-45202, Type I, Classes 1, 2, and 3.

Representative surface preparations are:

   I. a. Degrease
      b. Immerse for 10 minutes at 60 to 71C (140 to 160F) in alkaline detergent solution with the following composition:
         Water–95 parts by wt
         Sodium metasilicate–2.5 parts by wt
         Trisodium pyrophosphate–1.1 parts by wt
         Sodium hydroxide–1.1 parts by wt
         Nacconal NR (Aniline Division, Allied Chemical Co.)–0.3 part by wt
      c. Rinse thoroughly and dry, at not over 60C (140F).

  II. a. Degrease
      b. Immerse for 10 minutes at 71 to 88C (160 to 190F) in the following solution:
         Water–4 parts by wt
         Chromic acid (CrO_3)–1 part by wt
      c. Rinse thoroughly and dry, at not over 60C (140F).

 III. a. Degrease
      b. Immerse for 5 to 10 minutes at 63 to 79C (145 to 175F) in the following solution:
         Water–12 parts by wt
         Sodium hydroxide–1 part by wt
      c. Rinse thoroughly in water
      d. Immerse for 5 to 15 minutes at room temperature, in the following solution:

Water–123 parts by wt
Chromic acid–24 parts by wt
Calcium nitrate–1.8 parts by wt
    e. Rinse thoroughly and dry, at not over 60C (140F).

Plated Surfaces

Metals are usually plated as a corrosion-preventive measure or to retain surface conductivity for metals such as aluminum, which otherwise develops a dielectric film of aluminum oxide. Cleaning procedures involving acid-etch or abrasion are not advisable, since these would tend to remove the plating. If the surfaces are protected from contamination after the plating and cleaning process, they are usually clean enough for bonding. One method of obtaining improved bondability is to vapor hone or sandblast the surface before plating. This provides an increased surface-area as well as some mechanical benefits in increased shear-strength.

A plated surface which protects the base metal from corrosion does not necessarily improve adhesion or bond strength. Studies of adsorption on noble metals indicate that these adsorb less than metals such as copper.[16,17] Therefore, they might be expected to develop lower bond-strengths.

Gold and Silver Plating

Table 3-6 shows the results of bonding to unplated brass and to brass plated with silver or gold. Two different lots of a silver-filled epoxy adhesive are evaluated. The electrical resistivity of the bond is taken as an indication of oxidation at the unplated brass surface.

In every case resistivity is highest in the unplated brass specimens, which indicates a tarnished adherend surface. This high resistivity seems to correlate with highest bond-strength, and failure of the strongest bonds is always of the cohesive type. This indicates that an oxidized polar surface is more adsorptive and develops greater adhesion than a low-energy noble metal (nonoxidizing) surface.

Primers

Primers in common use as metal surface conditioners perform five different functions: (1) adhesive solutions compatible with the intended adhesive, applied immediately after cleaning, protect corrosion sensitive surfaces; they also help protect the more corrosion resistant metals if these are to be stored for any length of time; (2) adhesive resins are used as intermediate layers to modify the

Table 3-6.   Resistivity and Tensile Adhesion of Silver-Filled Epoxy Adhesive[a]

Lot Number	Adherend	Cure	Resistivity, ohm/cm	Tensile Adhesion, psi	Type of Failure
1	Brass[b]	3 hrs @ 149C (300F)	1820.0	3890	100% cohesion
	Brass, silver-plated	3 hrs @ 149C (300F)	0.56	2370	Partial failure in adhesion and of silver plating
2	Brass	16 hrs @ 135C (275F)	0.015	2090[c]	100% cohesion
	Brass, silver-plated	16 hrs @ 135C (275F)	0.0014	1440	100% adhesion to silver
	Brass, gold-plated	16 hrs @ 135C (275F)	0.0015	1760	95% cohesion

[a] Adhesive consisted of an epoxy resin cured with dicyandiamide, with approximately 65 percent silver-flake filling.
[b] The brass surface was cleaned by abrading with crocus cloth and wiping with clean methyl-ethyl ketone immediately prior to bonding. Silver- and gold-plated surfaces were plated about 4 hours before bonding and were wiped with clean methyl-ethyl ketone immediately prior to bonding. Bond thickness was approximately 0.010-inch and was measured for each specimen.
[c] Average of four values; all other averages represent five values.

properties of the bond; (3) adhesive resins which develop tack at room temperature or at elevated temperatures are used as processing aids for holding or positioning bonded parts; (4) corrosion inhibition; (5) multifunctional monomeric molecules are used for bond improvement. One primer could conceivably be used to serve two, three, or even all of the first four functions; the fifth function is usually independent of the first four.

The protective type of primer is widely used in the aircraft industry with epoxy, nitrile rubber phenolic, nylon epoxy, and epoxy phenolic adhesives. These adhesives quite often form non-tacky or slightly tacky films which are activated by heat and pressure to assume a fluid or semi-fluid state. The primers are applied as soon as possible after the metal surfaces are cleaned; then they are air and/or oven dried so that the metal parts can be handled and stored conveniently. The usual precautions regarding protection from environmental contamination are necessary.

During assembly the primed surfaces are usually activated (as the adhesive) by heat and pressure and are fused together with the adhesive to cure as one cohesive assembly. A primed surface should promote improved wetting and better adhesion than a similar, unprimed surface.

Because of the proprietary nature of adhesive formulations, the primers also are proprietary in nature and are made to match the adhesives. An adhesive manufacturer may sometimes recommend a primer made by a company which specializes in primer development, but this is not usual. When primers not specifically recommended by the adhesive manufacturer are used, preliminary evaluation is highly advisable. The manufacturer may or may not recommend the use of a primer, depending on the nature and economics of the application. A new production application where the bond is not as critical as it is, for instance, in the aircraft industry, might emphasize speed of application and cure with minimum impedance to the assembly line requirements. A primer might add an extra, unneeded, time-consuming step.

A second type of primer is designed to modify the characteristics of the adhesive. One of the more common combinations is an elastomeric primer with an epoxy adhesive. An unmodified epoxy adhesive has a very low peel strength, usually of the order of 5 pli or less. Because of this, its use in joint designs having peel or cleavage stresses may be inadvisable. This situation is improved by coating the metal surfaces with a nitrile rubber resin, a vinyl butyral resin, a neoprene primer, a polysulfide adhesive sealer, or any compatible rubbery material. Preliminary evaluation of such a bond is important because the final results are dependent on primer thickness and type of cure as well as compatibility. This combination of an epoxy adhesive with an elastomeric primer allows one to exploit the application and cure advantages of an epoxy adhesive while obtaining a tougher bond. Table 3-7 illustrates some of these advantages. The specimens tested consisted of an epoxy glass prepreg laminated to maraged steel. With the exception of the nitrile rubber phenolic No. 2 in the cleavage specimens, all the applications of primer improved the bond strength.

A third type of primer, which retains tack at room temperature or develops tack at elevated temperatures, is used for holding the adhesive film or adherend in place during assembly.

Table 3-7. Effect of Primers on Strength Properties of Glass/Epoxy-Laminations[a] to Maraged Steel[e]

Primer Type	Temperature		
	−54C (165F)	Ambient	149C (300F)
	Cleavage, pli		
None	70	63	32
Nitrile-rubber phenolic No. 1[b]	106	88	55
Nitrile-rubber phenolic No. 2[c]	57	53	45
Vinyl phenolic[d]	81	84	50
	T-Peel, pli		
Controls (no primer)	10	5.9	2.9
Nitrile-rubber phenolic No. 1	18	18.	11.
Nitrile-rubber phenolic No. 2	17	13.	10.
Vinyl phenolic	21	17.	20.
	Tensile Shear Strength, psi		
Controls (no primer)	1500	1090	1580
Nitrile-rubber phenolic No. 1	1920	1400	1510
Nitrile-rubber phenolic No. 2	2100	1950	1740
Vinyl phenolic	2220	1770	1080

[a] Scotchply XP-109-26S from 3M Co.
[b] BR-238 from Bloomingdale Department, American Cyanamid Co.
[c] EC-2174 from 3M Co.
[d] FM-47 (solution) from Bloomingdale Department, American Cyanamid Co.
[e] All specimens cured for 6 hours at 177C (350F).

The tacky condition which sometimes occurs at room temperature during the drying period lasts for a limited time, during which assembly must take place. Tackiness which is developed at elevated temperatures is much stronger and has much more stringent requirements. One example is found in the assembly of helicopter rotor blades. These are composed of a variety of internal ribs and spars in addition to the outer skin. After the mating surfaces are primed and assembled with the adhesive film in place, electrically-heated clamps are used to develop tack to weld the parts together in a fixed, predetermined position. After the assembly has cooled to room temperature, the clamps are removed. The parts are then firmly in position, held by the high-strength tack developed at the elevated temperature. The entire assembly is bagged and placed in an autoclave, and a vacuum is pulled on the bag. Pressure and heat are applied in the autoclave to cure the adhesive. Without the use of the tacky primer to hold the assembly firmly during the bagging and subsequent handling, much more complicated tooling would be required. Here again, the manufacturer's recommendations must be followed closely to get the full benefit from the primer.

The fourth use of primer as a corrosion inhibitor is essentially related to the corrosion protection primers as used in coatings. This type is described in detail by Krieger.[18]

A fifth use of primers for bond improvement—by means of multifunctional monomeric compounds—ties in with the mode of bond failure discussed in the previous chapter. Where bond failure is the result of preferential adsorption of water at the adherend surface accompanied by desorption of the adhesive, primers are used to condition the surface, probably by reacting chemically with chemisorbed surface films or polar groups, so that water is no longer the preferred adsorbate. The adhesive then competes on much more favorable terms for bonding sites.

Vincent describes a chemical treatment for an aluminum surface which improves the adhesion of polyethylene and also improves the resistance of the bond to water.[19] The function of silanes in glass surface treatments is described by Bascom with some speculation as to whether they act as chemical couplings or simply form a hydrophobic coating.[20] These types of primers have resulted in significant bond improvement with three different adhesive systems.

Table 3-8 shows the effect of a primer used with a silicone room-temperature vulcanizing system. Without the primer, negligible adhesion resulted. With the primer, T-peel strengths of 64 pli or

**Table 3-8. Effect of Primer[a] on T-Peel Strength of RTV
Silicone-Rubber Sealant-Bond[b] to Aluminum**

Surface	Environmental Exposure	T-Peel, pli	Type of Failure
Acid etch, (MIL-A-9067), no primer	None	Negligible	Adhesive
Acid etch, primed	None[b]	75	100% cohesive
Acid etch, primed	20-day humidity[c]	67	100% cohesive
Acid etch, primed	24 hrs @ 260C (500F)	64	100% cohesive
Acid etch, primed	24 hrs @ 316C (600F)	64[d]	100% cohesive

[a] Primer A 4094 from Dow Corning Corp.
[b] RTV 93-067 from Dow Corning Corp., cured 24 hours at 71C (160F) ± 3C (± 5F).
[c] Relative humidity > 90 percent with temperature cycle of 32C (89F) 24 hours + 65C (149F) 24 hours.
[d] Average of four values. All others, average of five values.

better resulted. These bonds were very resistant to high humidity and high temperature. While the materials used are proprietary in nature, indications were that the primer was a silane type with chemically active terminations.

Table 3-9 shows the result of using a primer with a polyurethane adhesive. Without the primer, T-peel strength eventually decreased from a high value of 247 pli to a very low 4.2 pli after exposure to high humidity. The primed surfaces bonded with the same adhesive retained most of the original strength, with a low of 188 pli after humidity exposure. The bond failures with the primer were primarily cohesive in nature. In these two instances, the only difference in the preparation of the durable and the nondurable bonds was the use of primers. Low-molecular-weight material was undoubtedly present in both the primed and unprimed specimens, especially in the silicone resin. The primed bonds were independent of these potentially weak boundary layer materials.

Table 2-1, taken from Reference 21, shows the effect of a silane finish in improving the wet strength retention of glass/resin laminates. The specificity of the silane action is shown by the fact that the silanes do not give equally good results with all the resins. This implies that chemisorbed bonds dependent on the resin silane reactions are responsible for the improved performances.

Similar silane coupling agents used as primers are described by Plueddemann for use with both adhesives and coatings.[22] These materials will probably find their greatest use in exterior grade paints, since these are specifically designed to protect substrates from an aqueous atmospheric environment.

Table 3-9. Effect of Primer on T-Peel Strength of Polyurethane Bond[a]

Surface Treatment	Primer Cure	Environmental Exposure	Peel Strength, pli[c]			Type of Failure
			High	Low	Avg	
Sulfuric/dichromate etch (MIL-A-9067)	...	None	315	130	247	Adhesive
No primer	...	15 days 95% RH	37	21	28	Adhesive
No primer	...	20 days SC 4451[d]	6.4	2.8	4.2	Adhesive
Sulfuric/dichromate etch Plus primer[b]	16 hrs @ 101C (215F)	None	300	160	222	100 percent cohesive
Plus primer[b]	16 hrs @ 101C (215F)	15 days 95% RH	268	170	208	95 percent cohesive
Plus primer[b]	16 hrs @ 101C (215F)	SC 4451[d]	230	160	188	100 percent cohesive
Plus primer[b]	16 hrs @ 101C (215F)	24 hrs in H$_2$O @ 88C (190F)	360	230	300	50 to 90 percent cohesive

[a] Formulation = Adiprene L100, E. I. du Pont de Nemours & Co., Inc. 100 pbw Cure: 40 hours @ 74C (165F)
     Dow Chemical P750
     MOCA, E. I. du Pont de Nemours & Co., Inc. 15 pbw Cure: 40 hours @ 74C (165F)
[b] Primer = Dayton Chemical Products XAB772. 6.6 pbw Cure: 40 hours @ 74C (165F)
[c] All values, average of five specimens.
[d] SC4451 = 20 days with humidity > 90 percent with temperature cycling between 65C (149F) and 32C (89F).

NOTE: The T-peel specimens consisted of two strips of .020-inch thick 2024 St aluminum with a sulfuric-acid dichromate etch, primed as noted above, with the polyurethane resin cast in between the strips. The specimens were pulled at 10 in./min.

Most of this chapter has been concerned with the importance of surface cleanliness as a preparation for adhesive bonding. However, though empirical in its beginnings, the use of chemically active adhesion promoters is here to stay. Understanding of the basic function of adhesion promoters or primers is making headway, and ultimately the use of primers will result in a major improvement in bond durability in both adhesives and coatings. Surface cleanliness will then be considered only the first step in surface preparation.

REFERENCES

1. Müller, E. W. "Field Ion Microscopy of Surface Structures on an Atomic Scale." *Symposium on Properties of Surfaces*, ASTM Special Technical Publication No. 340 (June 1963): 80–97.
2. White, M. L. *The Detection and Control of Organic Contaminants on Surfaces*. New York: to be published by Marcel Dekker, Inc., 1970.
3. Fjelseth, D. E.; Davis, D. M.; Jones, L. K.; and Schroeder, C. F. *Clean Assembly Practices Guide*. Sandia Corporation, October 1965.
4. Chessin, N., and Curran, V. "Preparation of Aluminum Surfaces for Bonding." In *Symposium on Structural Adhesive Bonding*, held at Stevens Institute of Technology, Hoboken, New Jersey, September 1965. New York: Interscience Publishers, 1966.
5. Bryant, R. W., and Dukes, W. A. "The Effect of Joint Design and Dimensions on Adhesive Strength." In *Symposium on Structural Adhesive Bonding*, held at Stevens Institute of Technology, Hoboken, New Jersey, September 1965. New York: Interscience Publishers, 1966.
6. Wegman, R. F.; Bodnar, W. M.; Bodnar, M. J.; and Barbarisi, M. J. "The Effects of Deionized Water Immersion of Prepared Aluminum Surfaces on Adhesive Bondability." *SAMPE Journal* (Society of Aerospace Materials and Process Engineers), October/November 1967: 35–39.
7. Twiss, S. B. *Adhesive Requirements for the Automotive Industry*, ASTM Special Technical Publication No. 360, June 26, 1963.
8. NASA Tech Brief 64-10142. Technology Utilization Officer, Goddard Space Flight Center, Greenbelt, Maryland, 20771.
9. Wegman, R. F. "Effects of Surface Preparation of Aluminum on Bonds Obtained with an Epoxidized Novolac Adhesive." *Picatinny Arsenal Technical Notes* FRL-TN-69, September 1961.
10. *Technical Bulletin 7101*. Erie, Pa.: Hughson Chemical Co.
11. Bloomingdale Department, American Cyanamid Co., Havre de Grace, Maryland.
12. *Recommended Practice for the Preparation of Metal Surfaces for Adhesive Bonding*, ASTM D2651. June 1968.
13. Keith, R. E.; Randal, M. D.; and Martin, D. C. *Adhesive Bonding of Stainless Steels*. NASA Technical Memorandum TM X-53574, February 16, 1967.
14. Reegen, S. L., and Ilkka, G. A. "Adhesion of Polyurethanes to Metals." In *Symposium on Adhesion and Cohesion*, General Motors Research Laboratories. American Elsevier Publishing Company, Inc. (1962): p. 163.
15. Keith, R. E.; Monroe, R. E.; and Martin, D. C. *Adhesive Bonding of Titanium and its Alloys*, NASA Technical Memorandum TM X-53313, August 4, 1965.

16. Fowkes, F. M. *Surface Chemistry.* Treatise on Adhesion and Adhesives. New York: Marcel Dekker, Inc. (1967): 380.
17. Daniel, S. G. Trans. Faraday Soc. *47* (1951): 1345.
18. Krieger, R. B. "Advances in Corrosion Resistance of Bonded Structures." *SAMPE Journal* (Feb./March 1969): 25–30.
19. Vincent, G. G. "Bonding Polyethylene to Metals," *Journal of Applied Polymer Science*, II (1967): 1553–1562.
20. Bascom, W. D. *Some Surface Chemical Aspects of Glass Resin Composites.* NRL Report 6140, August 10, 1964.
21. Sterman, S., and Toogood, J. B. "How to Promote Adhesion with Silicones and Silanes." *Adhesives Age.* July 1965.
22. Plueddemann, E. P. "Promoting Adhesion of Coatings Through Reactive Silanes." *Journal of Paint Technology,* Vol. 40, No. 516 (1968).

CHAPTER 4

Joint Design Test Specimens—
Stress Relations

Structural Bond

A structural bond is one which stresses the adherend to the yield point, thereby taking full advantage of the strength of the adherend. Thus, a structural joint designer taking into consideration adhesive characteristics should use this criterion as a guide. On the basis of this definition, a dextrin adhesive used with paper (e.g., postage stamps, envelopes, etc.) and which causes failure of the paper, forms a structural bond. The stronger the adherend (e.g., wood, glass-reinforced thermosetting plastic, or metal), the greater the demands placed on the adhesive. Thus, few adhesives qualify as "structural" for metals.

A further requirement for a structural adhesive is that it be able to stress the adherend to its yield point after exposure to its intended environment. An adhesive which decreases in strength significantly after extended exposure to an external environment would not be considered structural for that environment. Plywood is graded as either "interior" or "exterior," depending on the resistance of the adhesive to an exterior aqueous environment. Bonded metal structures have the same limitations; however, the environmental resistance of a bonded aircraft panel is sometimes upgraded with an edge seal, usually a polysulfide adhesive sealant, which protects the structural bond. Figures 2-2 and 2-3 illustrate the difference in behavior under severe conditions of two adhesives normally considered to be structural.

Stress Relationship

A third consideration in a structural bond is the adhesive/joint stress relationship. A relatively hard, brittle adhesive such as an epoxy may be very good in a joint designed for shear stresses with negligible peel components but very bad where peel forces are present. This type of adhesive would be structural in an assembly having only shear stresses or shear-tensile combined stresses. Where a structural bond may include peel stresses, then a tougher, high-peel adhesive such as a nylon epoxy or nitrile rubber phenolic should be used.

Of course, not all metal-to-metal bonds are structural. The primary consideration for nonstructural bonds is that the adhesive bond be good enough to withstand the intended service condition with a reasonable factor of safety, without necessarily having the ability to stress the metal to its yield point. In a nonstructural bond the cost and processability of the adhesive may become as important a factor as performance.

BOND EFFICIENCY

Bond efficiency—as determined by thickness of adherend, amount of overlap, and type of adhesive—is described in detail by Lunsford.[1] Adhesives of different modulus will differ in shear-stress characteristics depending on ability to relieve stresses across the bond area. In a critical structural assembly such as an airplane or missile, where weight saving and bond efficiency go hand-in-hand, it becomes very desirable to determine adhesive characteristics as accurately as possible. An empirical approach to characterizing an adhesive is to plot the l/t ratio against shear stress for each adhesive being considered for the structural bonds.

l/t is the ratio of length of overlap to adherend thickness. A complete story for each adhesive would require a series of l/t ratios versus shear stress for three different temperatures, that is, one curve for room temperature and one curve each for the high- and low-temperature extremes expected for the structure being designed. Each curve would apply not only to the specific adhesive but also only to the one adherend and temperature. A representative set of curves for a nitrile rubber adhesive with a double lap-shear specimen is shown in Fig. 4-1.[1]

Joint Design

Once an adhesive is characterized with regard to strength, as in Fig. 4-1, the simplest approach to designing a double-lap-shear joint using the same aluminum alloy and sheet thickness would be to select an overlap length which would result in a failing load that

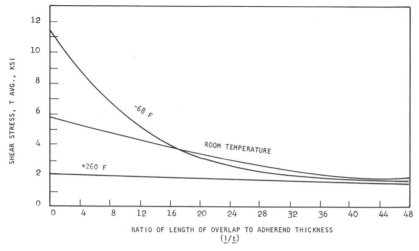

Fig. 4-1.   Nitrile rubber phenolic adhesive—double overlap specimens.

would stress the aluminum sheet to its elastic limit or yield point. Given an aluminum 0.025-inch thick, with a yield point of 60,000 psi and an overlap bond of 1 square inch, it would take a load of only 1500 pounds to reach the yield point. Thus, with an adhesive as in Fig. 4-1, which under the worst conditions (260F) achieves almost 2,000 psi with an l/t of 20, a ¾-inch overlap would satisfy the requirement that the bond stress the adherend to the yield point.

For a more exact method of determining the overlap length, Lunsford lists the following steps:

1. Obtain a trial overlap length by dividing the allowable shear strength corresponding to an l/t of 16 (Fig. 4-1) into the load to be transferred across the joint (the ultimate being that load which will stress the metal to the yield point).
2. Using this trial overlap and the actual adherend thickness, calculate the corresponding l/t.
3. Obtain the allowable shear stress (Fig. 4-1) corresponding to the l/t value calculated in step 2.
4. Using the allowable shear stress from step 3, calculate the second trial overlap length.
5. Repeat steps 2, 3, and 4 until the corrections to the required overlap length are negligible.

A separate l/t-shear stress relationship must be determined for each temperature, adhesive, adherend, and shear joint type being considered, since stress relationships will change if any one of these four variables is changed.

The curve changes noted in the Fig. 4-1 curves for the different temperatures are probably due to modulus changes with temperature; this emphasizes the importance of modulus as it affects stress distribution in an adhesive joint. The empirical approach to analyzing an adhesive as just described has the virtue of evaluating a reasonably representative joint with all its inherent flaws and thereby provides a realistic set of values on which to base a structural design.

Adhesive Moduli

Table 4-1[2] gives the moduli for various types of adhesives. Comparison of these modulus values with the peel and cleavage values in Table 3-7 gives some idea of bond property dependence on modulus. The prepreg bond to unprimed metal, that is, unmodified epoxy (highest modulus in Table 4-1) to metal, resulted in the lowest peel and cleavage values. The nitrile rubber primer with the lowest modulus gave the greatest improvement in T-peel strength, especially at the elevated temperature where the primer modulus would be lowest.

Table 4-2 shows the relationship between applied load, bond area, and sheet stress in a double lap-shear specimen.[3] Here even the lowest overlap area stresses the metal to the yield point. In this case, increasing the overlap area gains very little in structural strength. The increase, with overlap, in applied load and sheet stress may possibly be due to the longer time taken to complete rupture of the specimen with the longer overlap.

Single and double lap-shear joints are probably the most widely used in structural assemblies because they facilitate designing a joint which will stress a metal to the yield point. Because of this, shear specimens have been most thoroughly analyzed both empirically (as above) and theoretically. De Bruyne[4] and Golland and Reissner,[5] were among the first to recognize the importance of analyzing stress distribution in the bond area and in the adhesive/adherend system. More recently, Lunsford has done a theoretical analysis using a torsional shear specimen.[6]

A theoretical analysis is beyond the scope of this chapter but the references just mentioned would give one a good introduction to the theoretical aspects of the problem.

Table 4-1. Modulus of Rigidity of Several Adhesives

| | Torsion joints | | | | | Lap joints | | |
| | Tooley | Kuenzi | Lunsford | Lunsford | Kuenzi and Stevens | Wan and Sherwin | Broding | Eickner |
Adhesive	Modulus of Rigidity, psi					Modulus of Rigidity, psi		
Neoprene phenolic	...	...	...	...	...	2,680	...	...
Epoxy per MIL-A-8623, Type II	...	180,000	...	...	...	...	...	26,311
Epoxy phenolic	...	...	...	...	160,000	...	...	...
Nitrile phenolic	...	1,530	...	...	...	...	...	8,026
Nitrile phenolic	3,000	...	...	...	5,520	...	...	...
Nylon epoxy	...	...	...	...	49,300	...	...	...
Vinyl phenolic	150,000	...	154,600	...	117,000	...	...	...
Nylon epoxy	...	...	...	...	64,100	...	...	...
Vinyl phenolic	...	...	...	...	...	...	37,000	...
Nylon epoxy	...	...	...	31,000	...	...	...	...
Nitrile phenolic	...	...	...	6,850	...	...	...	...

Table 4-2. Applied Load versus Stress and Elongation in a Structural Bond[a,b]

No.	Applied[c] Load (lbs)	Bond Area, sq in.	Bond Stress, psi	Sheet Area, sq in.	Sheet Stress, psi	Elonga- tion, in.	Elonga- tion, percent
1	3040	0.75	4050	0.050	60,800	. . .	. . .
2	2940	0.75	3920	0.050	58,800	. . .	. . .
3	3040	0.75	4050	0.050	60,800	. . .	. . .
4	2990	0.75	3990	0.050	59,800	. . .	. . .
Avg	3000	0.75	4000	0.050	60,000	0.360	5.75
6	3230	1.00	3230	0.050	64,600	. . .	. . .
7	3280	1.00	3280	0.050	65,600	. . .	. . .
8	3210	1.00	3210	0.050	64,200	. . .	. . .
9	3140	1.00	3140	0.050	62,800	. . .	. . .
Avg	3215	1.00	3215	0.050	64,300	0.53	8.85
11	3400	1.50	2266	0.050	68,000	. . .	. . .
12	3295	1.50	2196	0.050	65,900	. . .	. . .
13	3345	1.50	2230	0.050	66,900	. . .	. . .
14	3345	1.50	2230	0.050	66,900	. . .	. . .
Avg	3346	1.50	2230	0.050	66,920	0.845	12.1

[a] The specimen was a double lap-shear type prepared from 2024-T3 aluminum strip 0.025-inch thick by 1-inch wide. The aluminum surface was treated with a sulfuric acid-dichromate solution per MIL-A-9067.
[b] The adhesive was a polyvinyl-butyral-phenolic cured at 177C (350F) for 1 hour at 50-psi pressure.
[c] Rate of head-travel was 0.050 ipm.

Joint Design

A joint should be designed to take advantage of the best properties of the adhesive and, if possible, the strength of the adhesive joint should be of the same order of magnitude as the strength of the metal. This is especially true in aircraft structures since overdesign in terms of bond strength would incur a weight penalty. In addition, with most adhesives of medium to high modulus, peel and cleavage forces should be minimized. For example, a pure butt joint in a metal-to-metal bond (see Fig. 4-2) places the adhesive in a very unfavorable position since:

1. The bond is a great discontinuity in the strength of the assembly. Assuming the bond has a strength of 5000 psi, it has less than 1/10 the tensile strength of aluminum and less than 1/20 the tensile strength of the usual steel alloys. Thus a pure butt-joint represents a serious flaw in the continuity of the assembly's strength.

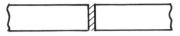

Fig. 4-2.   Pure butt joint.

2. A butt joint is sensitive to side loading, thereby subjecting the bond to cleavage forces.

If the metal is too thick to design as a simple lap-shear joint, then the simple butt-joint could be improved by redesigning as a tongue-and-groove joint or a double strap reinforced butt joint (see Fig. 4-3).

Butt Joint Variations

In the tongue-and-groove and double lap, the cleavage effect of side loading is eliminated because the metal bears practically the

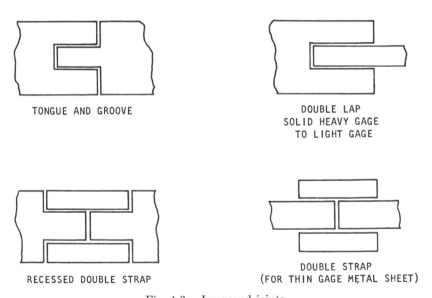

TONGUE AND GROOVE

DOUBLE LAP
SOLID HEAVY GAGE
TO LIGHT GAGE

RECESSED DOUBLE STRAP

DOUBLE STRAP
(FOR THIN GAGE METAL SHEET)

Fig. 4-3.   Improved joints.

full effect of the side load. The double strap variations tend to minimize side-load effect. In all the above joints the combined tensile and shear surfaces included in the bond have been doubled or tripled, depending on the amount of overlap, so that the strength discontinuity is minimized.

Single-strap variations of the above are shown in Fig. 4-4.

Shear Joints

Variations of the double lap-shear would include the solid double-lap (Fig. 4-5), which was the double lap used for obtaining the data in Table 4-2 and the recessed double-lap. Recessing results in a more cleanly-finished part and protects the edges.

Double lap-shear designs are very often used with honeycomb core construction in aircraft panels. In his detailed analysis of

RECESSED SINGLE STRAP   SINGLE STRAP

Fig. 4-4.   Single strap variations.

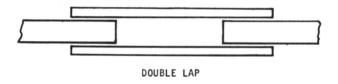

DOUBLE LAP

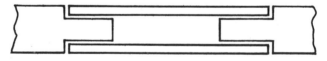

RECESSED DOUBLE LAP

Fig. 4-5.   Variations of the double lap shear.

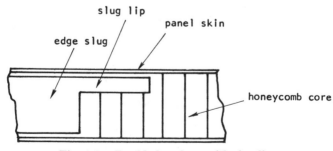

Fig. 4-6.   Double lap shear with slug lip.

double lap-shear designs, Lunsford[1,6] describes a double lap-shear with a slug lip which facilitates transfer of core shear loads to the edge slug (see Fig. 4-6).

Double lap-shear designs have the advantage of having stresses uniform around the center of load application; single lap-shear joints are nonuniform in this respect, which contributes to uneven stress distribution (see Fig. 4-7). This type of deformation can be minimized with the use of a scarf joint or beveled overlap (see Fig. 4-8). The scarf joint keeps the axis of applied load centered with respect to the joint, but the joint itself is difficult to prepare with regard to alignment and pressure application. The beveled lap minimizes the transfer of edge stresses to the bond because of the feathered edges and is easier to prepare than the scarf joint. The beveled edge would be applicable to any shear joint with an exposed corner, as in Figs. 4-4, 4-5, and 4-7.

Figure 4-9 illustrates effective variations of the single lap-shear joint which require machining but which would allow practical processing.

Corner Joints

Corner and angle joints are essentially variations of the butt and shear types already described. However, a corner joint, because of its lack of symmetry, is inherently exposed to side loading or cleavage. Therefore, the most successful joint is the one of tongue-and-

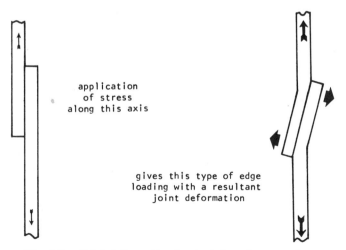

application
of stress
along this axis

gives this type of edge
loading with a resultant
joint deformation

Fig. 4-7.   Joint deformation in single lap shear specimen.

groove design, since this places the side loading on the metal which, as has been pointed out, is 10 to 20 times stronger than the adhesive. This type is practical also, since it gives positive alignment with a minimum of jigging (Fig. 4-10).

In strap-supported variations (see Fig. 4-11), it is immediately evident that the bond would have to support the cleavage loads.

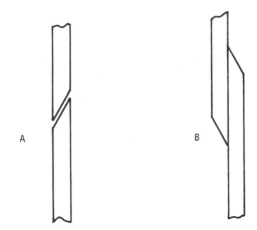

Fig. 4-8.   (a) Scarf joint, and (b) beveled overlap.

Fig. 4-9.   (a) Double butt lap, and (b) double scarf lap.

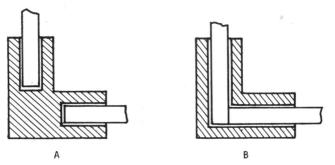

Fig. 4-10.   (a) Slip joint, and (b) right-angle support.

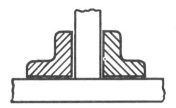

Fig. 4-11.   Strap-supported joint.

Thus, a tough, high-peel-strength adhesive would be preferable to a low-peel-strength adhesive in these designs.

Tubular Joints

In tubular joints (Fig. 4-12) the designs more nearly favor pure shear. Cleavage loads are eliminated because the metal encloses the bond. But processing may be complicated since, in pushing one cylinder into another, as in Fig. 4-12D and E, an adhesive applied to both mating surfaces tends to be scraped away and air is pulled into the bond. The amount of actual contact is uncertain. This is partially overcome in a butt lap-joint such as Fig. 4-12B, where the adhesive in the corner areas is subject to pressure resulting in a positive filling action. The tubular joints which would insure a positive pressure on the adhesive and a complete fill are the tapered or scarf joint (Fig. 4-12A) and variations thereof (Figs. 4-12C and 4-12F).

If the tubular joint is subject to sufficient tensile force to deform the metal elastically, a design such as Fig. 4-12G is desirable. Elastic deformation due to tensile loads would tend to cause "necking" in the tube,[7] thereby putting a compressive load on the bond. This would favor a pure shear-stress. In all of the other joints, elastic deformation in the tubing would induce peel or cleavage stresses.

In straight-sided tubular joints a positive adhesive fill can be achieved by injecting the adhesive into the joint after assembly, as shown in Fig. 4-13. This does require close tolerances in a design which would confine the adhesive so that it is forced around the tube from the injection hole to the exit hole. Clearances would have to be such that the adhesive would be forced to flow the full length of the external or the internal sleeve as in Fig. 4-12G. Clearances of 0.005-0.006 inch have been found effective for highly filled epoxy adhesives as specified in MIL-A-8623.[8] If tubular joints of greater

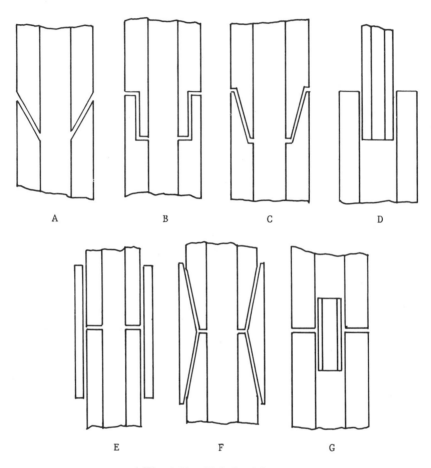

Fig. 4-12.   Tubular joints.

than 1-inch diameter are jointed in this manner, it would be necessary to use more than two holes. The procedure would then be to
inject the adhesive into one hole until it flowed to the two adjacent
holes and then continue around the joint until the joint was completely filled.

TEST SPECIMENS

A bonded assembly may consist of many different joints subject
to all conceivable stresses and combinations of stresses. A materials

Adhesive

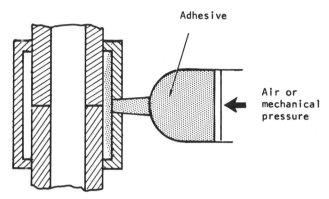

Air or
mechanical
pressure

Fig. 4-13.   Straight-sided tubular joint.

engineer specializing in adhesives must have access to accurate information on the stresses to be expected in the joint and on the properties to be expected from the wide variety of adhesives now available for structural bonding. The stresses to be anticipated resolve themselves into three basic types: shear, tensile, and peel. A variation of peel which involves localized stresses in heavy gauge, high-modulus adherends is called cleavage. Test methods and specimens have been standardized and are available to measure each of these stresses for adhesive-bonded specimens. This makes it possible for different laboratories to evaluate adhesives and obtain comparable results.

TEST METHODS AND ASTM[9]

When one discusses test methods, the American Society for Testing and Materials (ASTM) inevitably becomes part of the discussion. ASTM, which originated in the United States, has become an international organization devoted to the promotion of knowledge of materials and the standardization of specifications and methods of testing. ASTM brings together the knowledge and experience of general interest groups, producers, and consumers to write test methods and specifications on a nonpartisan basis so that they represent the best balance of all the interests involved. The test methods and specifications sometimes originate as industry or industrial organization standards before being rewritten by ASTM to represent a broader point of view.

Once written, the standardized test methods and specifications are used not only in the member countries of the Americas, but through USASI (United States of America Standards Institute) they are presented to the ISO (International Standards Organization) for consideration and adoption. Thus, through these standards, the ASTM technical committees help promote international understanding and cooperation. Standardization of test methods also encourages exchange of ideas and goods.

One of the technical committees involved in this far-reaching process of standardization is the ASTM committee D-14 on adhesives. D-14 was organized in 1944 and has been instrumental in writing up more than 60 test methods, specifications, and recommended practices as well as a set of definitions of terms relating to adhesives. These methods cover a wide variety of tests ranging from lap shear to "Susceptibility to Attack by Roaches of Dry Adhesive Films," and are to be found in part 16 of the 1968 ASTM Book of Standards. This review will only touch on some of the basic test methods described in part 16 and on some which have not yet been considered by the D-14 committee.

Shear Tests—Tensile Test

Pure shear stresses are those which are imposed parallel to the bond and in the plane of the bond. The most widely-used specimen to measure shear is illustrated in Fig. 4-7. It does not represent pure shear, but it is practical and relatively simple to prepare, and it does give reproducible, usable values.

The preparation of this specimen and method of testing are described fully in the Standard Method of Test for Strength Properties of Adhesives in Shear by Tension Loading (Metal-to-Metal), designated as ASTM D1002-64. The description covers calculations for determining length of overlap, preparation of test specimens, and test procedure. Two types of panels for preparing multiple specimens are described. One type of panel is slotted so that it is not necessary to cut past the bonded joint to separate the individual specimens. The other type of panel is prepared by bonding two solid sheets of metal together with a ½-inch overlap. This panel must be carefully cut in order to avoid harming the adhesive bond when cutting past it.

Compression Test

A compression shear-test method (D905) is described which is intended primarily for wood, but could be adapted to metals. This is a block shear-test which would minimize cleavage or peel stresses.

D2295, Tensile Shear of Adhesives at Elevated Temperatures, is similar to D1002 but provides information on method of load application, thermocouple attachment, and heat-lamp arrangement.

D2182, Compression Disc Shear, is a complete test method in that it describes: (1) a relatively simple way to prepare a disc-on-metal-strip specimen, and (2) a shear jig including a thermostatted heater, which makes it possible to test at various temperatures without enclosing the whole test machine in an oven.

Tensile Tests

Pure tensile-adhesion tests are those in which the load is applied normal to the plane of the bond and in line with the center of the bond area.

D897, Tensile Properties of Adhesives, is one of the oldest methods in the book. The specimens and grips require considerable machining and, because of the design, tend to develop edge stresses during test. Because of these limitations, D897 is being replaced by D2095, Tensile Strength of Adhesives, Rod and Bar Specimens. The rod and bar specimens, prepared per D2094, are simpler to align and, when correctly prepared and tested, more properly measure tensile adhesion. A typical, rod tensile-adhesion specimen is shown in Fig. 4-14.

Peel Tests

Peel tests are designed to measure the resistance of adhesive bonds to highly localized stresses. Peel forces are therefore considered as being applied to linear fronts. The more flexible the adherend and the higher the adhesive modulus, the more nearly the stressed area is reduced to linearity. The stress then approaches infinity. Since the area over which the stress is applied is dependent on the thickness and modulus of the adherend and the adhesive, and therefore very difficult to evaluate exactly, the applied stress and failing stress are reported as linear values, that is, pli (pounds per linear inch). Probably the most widely used peel test for thin-gauge metal adherends is the so called T-peel test. The designation of the ASTM version is D1876. A distinctive characteristic of the T-peel specimen as described in D1876 is that all of the applied load is transmitted to the bond. This type of peel thus tends to give the lowest values.

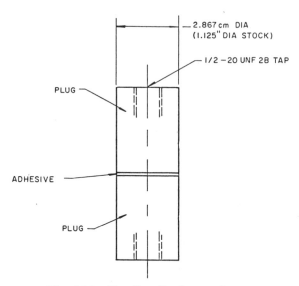

Fig. 4-14.   Tensile adhesion specimen.

With elastomeric adhesives such as castable polyurethane rubber, RTV silicones, and polysulfides, peel strength is dependent on bond thickness. Figure 4-15 shows the T-peel dependence on bond thickness for a polyurethane adhesive.[10] The elongation characteristics of these adhesives allow a greater area of the bond to absorb the applied load as the bond thickness increases.

Three types of metal-to-metal peel specimens and test jigs are shown in Fig. 4-16. The T-peel (ASTM designation D1876) is probably the most widely used and the simplest to prepare since it uses only one thickness of metal. The Bell peel, named after its developer, Bell Aircraft Corporation, is designed to be peeled at a constant radius around a 1-inch steel roll and thereby give more reproducible results. The metal-to-metal climbing-drum method, ASTM designation D1781, represents an attempt to achieve the same, constant peel radius by peeling around a 4-inch diameter rotating drum.

While the fixtures used with the Bell and drum-peel tests help to stabilize the angle of peel, the ideal of a fixed radius of peel is probably not realized because the high modulus of the metal tends to resist close conformation to the steel roll or drum. In both methods, considerable energy is used in deforming the metal so that

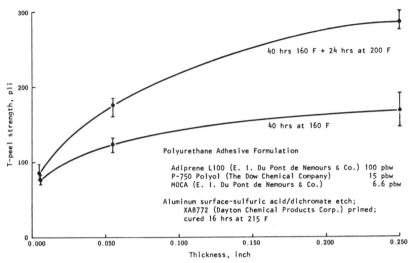

Fig. 4-15.   Effect of bond thickness and cure on T-peel strength.

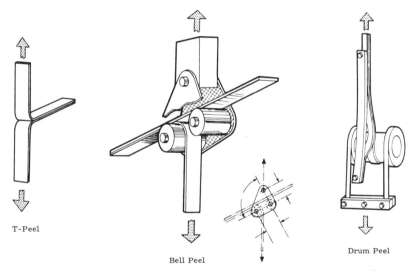

Fig. 4-16.   Three types of metal-to-metal peel specimens and test jigs.

they indicate higher values for a given adhesive than the T-peel method.

Representative figures obtained by the different metal-to-metal peel methods[11] are shown in Table 4-3. The difference in values resulting from the different test methods emphasizes the importance

Table 4-3. Results of Different Metal-to-Metal Peel Methods

Test Method	Test Temperature	Average Results, pli[a,b]
T-peel	−55C (−67F) 24C (75F) 82C (180F)	19 27 31
Bell peel	−55C (−67F) 24C (75F) 82C (180F)	52 60 60
Metal-to-metal climbing-drum peel	−54C (−65F) 24C (75F) 82C (180F)	42 66 61

[a] All values, average of six or more specimens.
[b] Adhesive is a modified epoxy, 0.060 lb/sq ft film weight, made by Blooming-dale Department of American Cyanamid Co., designated FM 123-2.

of giving complete details as to method, materials, and conditions of testing when reporting the test result.

The climbing-drum method was developed primarily for use with honeycomb-core sandwich panels to measure the resistance to peel of the sandwich skin.

Cleavage Tests

Cleavage is a variation of peel in which the two adherends are classified as rigid. The load is applied normal to the bond area at one end of the specimen, as shown in Fig. 4-17. The ASTM designation is D1062. This describes, in detail, the specimen grips and method of test.

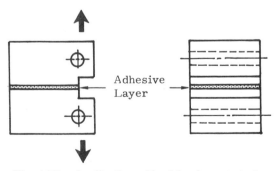

Adhesive
Layer

Fig. 4-17.   Application of load in cleavage test.

Creep Tests

Quite often when a bonded structure is subject to a permanent load in service, especially in the presence of vibration, the adhesive's resistance to creep is an important characteristic. Two ASTM methods are available for evaluating this property: D2293, Creep Properties of Adhesives in Shear by Compression Loading (Metal-to-Metal), and D2294, Creep Properties of Adhesives in Shear by Tension Loading (Metal-to-Metal). These two methods use lap-shear specimens with springs to apply the compression or tension load.

Fatigue Tests

Static strength tests perform a necessary function in that they make it possible to screen and select adhesives for most bonded applications with a fair degree of accuracy. When they are combined with representative stresses and environment, they diminish even more the uncertainties involved in selecting adhesives for critical applications.

One of the most rigorous conditions is that of intermittently applied stress or fatigue. ASTM D-14 has not yet written up a test method for fatigue possibly because fatigue testing of a bonded joint is likely to cause failure of the metal rather than the bond. However, fatigue testing of bonded assemblies is important, if only to evaluate a bonded assembly as compared with a riveted assembly.

Probably the most dramatic test for the effect of fatigue on a bonded assembly took place during the investigation, conducted in England, of the cause of failure of the first commercial jet planes, the de Havilland Comet I series.[12] Because failure due to fatigue in the aluminum skin was suspected as a cause of the disasters, one of the Comets was immersed in a tank full of water and, with hydraulic pressure simulating the pressure cycle experienced by a plane flying at altitudes from 35,000 feet to sea level, was subjected to the equivalent of 9,000 flying hours. At this time, fatigue cracks appeared on the body of the plane. These cracks, along with other corroborating evidence, pinpointed the reason for the failures and eliminated doubts as to the durability of adhesively-bonded sections of the Comet.

Fatigue tests are not usually carried out on the scale necessitated by the Comet disasters. Initially, tests are run on the usual specimens with the imposed cycle designed either for a constant load or constant deflection. The equipment for this is described in detail in ASTM D671, Tests for Flexural Stress (Fatigue) of Plastics.

Impact Tests

Impact tests bring into the testing picture the ability of an adhesive to attenuate or absorb forces applied in a very short time interval. Essentially, these tests measure an adhesive's rate-sensitivity to an applied load. Some machines used gravity to accelerate a given load which strikes the test specimen. The amount of energy absorbed in breaking the bond is a measure of bond strength. ASTM Test for Impact Strength of Adhesives (D950) describes a pendulum method for applying an impact load to a shear specimen. The test value is given as foot-pounds of energy absorbed in failing the bond of a 1-square-inch shear specimen.

Information on impact testing machines may be found in the Methods for Notched Bar Impact Testing of Metallic Materials (ASTM Designation, E23).

Another variation of the gravity-impact method uses a series of weights dropped on the test specimen. The failing load would be given as weight times distance dropped.[13]

Other, more sophisticated machines use compressed air to decrease the time of load application to as little as 10^{-5} second.

Tests for Lightweight-core Panels

Panels with metal skins and lightweight cores are widely used in the aircraft and construction industries. In the aircraft industry, the lightweight core is usually resin glass or metal honeycomb; in other fields, it may be paper honeycomb, balsa, or some other lightweight material.

The lightweight core is a completely bonded structure, and tests relating to the materials composing it (except for the climbing-drum peel test, already discussed) come under the jurisdiction of ASTM Committee C-19 on Structural Sandwich Construction.

A list of test methods follows:

ASTM Designation	Title
C 364-61	Compressive Strength, Edgewise, of Flat Sandwich Constructions
C 365-57	Compressive Strength, Flatwise, of Sandwich Cores
C 363-57	Delamination Strengths of Honeycomb Type Core Material
C 271-61	Density of Core Materials for Structural Sandwich Constructions
C 480-62	Flexure-Creep of Sandwich Constructions
C 393-62	Flexure Test of Flat Sandwich Constructions
C 481-62	Laboratory Aging of Sandwich Constructions
C 394-62	Shear Fatigue of Sandwich Core Materials
C 273-61	Shear Testing Flatwise Plane of Flat Sandwich Constructions or Sandwich Cores
C 297-61	Tension Test of Flat Sandwich Constructions in Flatwise Plane
C 366-57	Thickness of Sandwich Cores, Measurement of
C 272-53	Water Absorption of Core Materials for Structural Sandwich Constructions

While standard test methods and specimens are developed to insure that values from different laboratories and industries can be correlated, specific applications and service conditions may require variations from established procedures. In using tests and conditions which vary from the standard, it should be kept in mind that exact reporting of test conditions and results is necessary to make them meaningful.

REFERENCES

1. Lunsford, L. R. "Design of Bonded Joints." In *Symposium on Adhesives for Structural Applications*, edited by M. J. Bodnar, 9-14. New York: Interscience Publishers, 1962.
2. Kutscha, Dieter. *Mechanics of Adhesive Bonded Lap Type Joints Survey and Review*. ML-TDR-64-298, FPL Madison, Wisconsin, December 1964.
3. DeLollis, N. J. "Structural Adhesives." Paper read at ASTME Creative Manufacturing Seminar, Cleveland, Ohio, January 30, 1969.
4. DeBruyne, N. A. "The Strength of Glued Joints," *London Aircraft Engineering* 16 (1944): 115.

5. Goland, M. and Reissner, E. "The Stresses in Cemented Joints," *Journal of Applied Mechanics*, II, No. 1. (1944): A17.
6. Lunsford, L. R. "Stress Analyses of Bonded Joints." In *Structural Adhesives Bonding*, edited by M. J. Bodnar, 57–73. New York: Interscience Publishers, 1966.
7. Popov, E. P. *Mechanics of Materials*. Englewood Cliffs, N. J.: Prentice-Hall, Inc., 1952.
8. DeLollis, N. J. "Structural Adhesives — Characteristics and Applications." In *Symposium on Adhesives for Structural Applications*, edited by M. J. Bodnar, 37-42. New York: Interscience Publishers, 1962.
9. ASTM — American Society for Testing and Materials, 1916 Race Street, Philadelphia, Pa., 19103.
10. DeLollis, N. J.; Montoya, O.; and Curlee, R. M. *Effect of Assembly Variables on Peel Strength of Adiprene Bonds*. Sandia Corporation, SC-DR-65-161, June 1965.
11. Values in table are given with permission of Bloomingdale Dept., American Cyanamid Co.
12. Dempster, Derek D. *Tale of the Comet*. New York: David McKay Company, 1960.
13. Cagle, C. V. *Adhesive Bonding Techniques and Applications*. New York: McGraw-Hill Book Co., 1968.

CHAPTER 5

Adhesive Materials—Structural

The adhesive materials to be discussed in this chapter are used primarily with metals. They bond well, with or without heat activation and pressure, to smooth impermeable surfaces and, compared to thermoplastic adhesives, are resistant to the action of solvents and temperature. They are synthetic in origin and are crosslinked. They do not replace any previous structural adhesives for metals, because structural bonds to metals did not exist until these adhesives were developed.

PHENOLIC COPOLYMERS—CHEMISTRY

The first structural-metal adhesives were composite polymers which had one synthetic resin in common—phenol formaldehyde. This polymer, discovered by Dr. Baekeland in 1909, was developed as an adhesive primarily for wood products.[1] It is a hard brittle resin which produces water as a condensation reaction by-product and as such does not bond well to metals. However, as elastomeric polymers were developed it was found that phenol formaldehyde and other phenol resins were almost the universal copolymerizing agents; they behave as both adhesion promoters and as additives which significantly improve temperature resistance.

Phenolic resins find wide use in structural adhesives as copolymers with four elastomeric resins. While these combinations have sometimes been considered primarily mechanical mixtures, the change in properties with degree of cure and the solvent resistance requirements listed in the various Military Specifications indicate that chemical cures must take place.[1]

The building blocks for these copolymers are:

Phenol dialcohol from
phenol and formaldehyde

Isoprene

The remaining hydroxyl group would then be available for similar reactions. The double linkage is also available for subsequent reactions with additives such as sulfur or metal oxides.

The reaction noted above could very easily take place with Chloroprene (neoprene) or nitrile rubber (Buna N).

The third commonly used structural adhesive is a phenolic-polyvinyl butyral type. Since the polyvinyl butyral contains up to 20 percent unconverted hydroxyl groups, the reaction here, as suggested by Martin and others, would be between the phenol dialcohol and the vinyl hydroxyl groups.[1, 2]

The fourth phenolic copolymer is with an epoxy resin. While the rubber and vinyl molecules of the first three copolymers impart a toughness to the phenolic resin, adding epoxy resin to the phenolic results in an adhesive characterized by low creep at temperatures up to 260C (500F), brittleness, and low peel strength.

Here, a possible reaction would be between the epoxy group and the phenol hydroxyl.

Epoxy   Phenol Dialcohol

$$- - - - C \overset{O}{\underset{H}{\diagup}} \underset{H}{C} + HO - CH_2 - \overset{OH}{\diagdown} - CH_2 OH$$

R

$$- - - - C \overset{OH}{\diagup} \underset{H}{C} - \underset{H_2}{C} - O - CH_2 - \overset{OH}{} - CH_2 OH$$

R

This reaction would also be accompanied by the usual crosslinking epoxy/amine reactions.

NEOPRENE ADHESIVES

Structural bonding of metals became a reality about 1942 when the neoprene phenolic adhesives were first developed.[3] These combined the heat resistance and strength of the phenolic resins with the toughness and stress-relief characteristics of rubber. The neoprene phenolics were available as solutions, which were brushed on the mating surfaces, dried before assembly, then assembled and cured at temperatures of 163C to 177C (325F to 350F) for 30 to 60 minutes with pressures of 200 to 300 psi. Bond strengths of up to 4,000 psi were achieved. This made possible metal-to-metal bonds which were stronger than riveted joints.

Brake Linings

One of the first applications with this adhesive was the bonding of automobile brake linings. While not strictly a metal-to-metal bond, it was structural in that it had to withstand intermittent high-shear loads at high temperatures. At the same time the bond was exposed to all the atmospheric extremes of temperature and moisture.

This high-strength version of the neoprene phenolic adhesives has been replaced by nitrile rubber phenolic adhesives, which process more easily. However, neoprene cements find wide use in industry for general nonstructural applications.[4]

One type meets the requirements of MIL-A-5092, Type II. It is an oil-resistant adhesive recommended for bonding neoprene rubber to neoprene and other materials. It has a limited shelf life of about three- to six-months depending on storage temperature. It is usually applied (1) as a wet-bonding adhesive where the adherends are assembled immediately after application; or (2) as a solvent-activated adhesive applied to both surfaces, then allowed to dry tack free (immediately before assembly, one of the mating surfaces is lightly coated with a compatible solvent or with the adhesive). The parts are assembled with slight pressure. One of the adherends bonded by these methods should be permeable or absorbent so that the bond can dry within a few hours. A bond strength of about 225 psi in shear can be expected under these conditions.

Heat Activation

Best results can be expected by heat activation where the coated surfaces are dried tack-free—perhaps with oven drying—at temperatures up to 74C (160F) for 30 to 60 minutes. The parts are then assembled and cured with heat and pressure—121C (250F) for 60 minutes and 50 to 100 psi. Shear strengths up to 1000 psi can be achieved with this procedure. Even impermeable materials, such as metals, can be bonded in this manner.

Contact Cement

The contact type, neoprene resin cements are among the most significant developments in solvent-based adhesives in the last few years. Until the development of contact adhesives, a big drawback in the use of solvent-based adhesives was that, for room temperature or intermediate temperature cures, the mating surfaces had to be assembled while wet. This meant a long, drawn-out, drying time while mechanical jigging was required to hold the assembly together as the bond built up strength. Also, as the solvent left the bond area, it was inevitable that voids formed in the adhesive layer.

With contact adhesives, it is possible to air-dry the coated mating surfaces until they are nontacky. In this state they will stick only to each other and not to any other surface. Assembly consists simply of bringing the surfaces together with sufficient pressure to achieve contact. The initial bond is then strong enough to hold the assembly with no additional support. Contact cements have found wide application in bonding a great variety of materials, both flexible and rigid, metallic and nonmetallic. They have minimized the need for expensive holding fixtures.

Since the mating surfaces cannot be slipped into position after contact, a simple technique has been developed to allow exact positioning before contact. The coated surfaces are dried until they cannot stick to clean paper, and are then assembled with a sheet of paper between them. Positioned as desired, the surfaces are held so they do not shift with respect to each other while the paper is withdrawn. Once the paper is withdrawn, the surfaces come into contact, and pressure is applied to establish a permanent bond.

Performance requirements for contact adhesives per Federal Specification MM M-A-130 are a shear strength of 150 psi immediately after bonding and 200 psi after a 7-day cure.

Contact adhesives can also be processed by forced oven drying with subsequent assembly using heat 121C (250F) and pressure (50 to 100 psi) for 30-60 minutes to give much higher shear strengths. Peel strengths with contact cements will run about 20 pounds per linear inch. Temperatures as low as 70C (158F) have been used successfully for heat activation.

A representative contact adhesive formulation is:

Ingredient	Pb wt
Neoprene stock	100
Antioxidant (Neozone A)	2
Magnesia	8
Zinc oxide	5
Phenolic resin	40
Toluene	400
Ethyl acetate	80
Water	1

Contact adhesives have the stability to exterior environment usually associated with neoprene; in addition, they seem to have overcome the problem of limited shelf-life usually associated with neoprene adhesives per MIL-A-5092, Type II.

VINYL PHENOLICS

The vinyl phenolics were first developed in England, about 1943, by Aero Research under the name of "Redux." In the Redux version, the adherend surface was coated with a phenolic liquid, the vinyl powder was spread on the wet phenolic surface, the excess was shaken off, and the parts were assembled and cured. In the United States, vinyl phenolic formulations soon became available as one-part solutions and films. Cure conditions are about 60 minutes at 177C (350F) with a pressure of about 15 to 50 psi.

A ratio of two parts polyvinyl butyral to one of phenolic resin would feature high, room-temperature peel. As the phenolic ratio is increased high-temperature strength is increased. Continued exposure to high temperature would tend to decrease peel strength but increase high-temperature shear.

The data in Table 4-2, illustrating the ability of a structural-shear bond to deform aluminum beyond its elastic limit, was obtained with a polyvinyl butyral phenolic adhesive.

Representative data for a commercially available polyvinyl-butyral-phenolic adhesive[5] is shown in Table 5-1.

Table 5-1. Representative Data for a Polyvinyl-Butyral Phenolic Adhesive*

Test Method	Controls	After 2 Hours at 177C (350F)	50 Hours Salt Spray	After 20 Days >95% RH
T-peel ASTM D1876	29[a,b]	6	33	38
Lap shear ASTM D1002	3380	3570	3360	3090

* Taken from Reference 6
[a] All values, average of five specimens
[b] T-peel specimen consisted of 1-inch wide strips of 0.025-inch thick aluminum, 5 inches long, bonded together for 4 inches of the length. Rate of loading: 10 in./min.

The vinyl phenolics feature good flow under moderate pressure (15 to 50 psi) with good peel characteristics. Their use would not be recommended where more than one bonding cycle is contemplated for an assembly, since their peel strength tends to decrease after exposure to high temperature. Operational temperature-range for this type would be from −240C (−400F) to +121C (+250F).

This type of adhesive still finds wide use in aircraft structures involving both skin- and node-bonding of aluminum honeycomb and metal-to-metal construction.

As shown in Table 3-7, vinyl phenolics are compatible with epoxies, and can be used as a primer prior to bonding with an epoxy adhesive to lend added toughness to the bond. Care should be taken to oven-dry the vinyl phenolic primer coating before assembly, otherwise the coating will outgas because of the heat needed to cure the epoxy. Outgassing results in a bubble filled, weak bond. Vinyl phenolics also find wide use in the fabrication of low-density core, structural sandwich panels.

Figure 5-1 serves well to illustrate how an adhesive (in this case a vinyl phenolic[5]), properly applied with a joint designed according to the right l/t ratio, can cause metal failure. In the case illustrated, metal failure occurs within the limits imposed by the nature of the adhesive and the thickness of the metal. With an l/t ratio of 30 and an aluminum thickness-range of 0.030 to 0.249, it is possible to cause metal failure over a total temperature-range from -51C (-60F) to $+127$C ($+260$F).

The adhesive modulus increases with decrease in temperature. Therefore, joint efficiency is decreasing, and the likelihood of bond failure increases. At the higher temperatures adhesive modulus is decreasing so that eventually the adhesive layer becomes weaker than the metal cross-section. This type of curve is characteristic of any organic adhesive. However, the optimum portion of the curve may shift higher or lower on the temperature spectrum as the adhesive type is changed; for example, the use of polyurethanes would shift the curve to the lower temperature regions while the use of epoxy phenolics would shift it to the higher temperature regions. Thus, service temperature-range is one of the factors which must be kept in mind when making an adhesive choice.

Outgassing

Phenolics in general tend to outgas, and the vinyl phenolics are no exception. Therefore they can be used with perforated-core honeycomb, but should be avoided in applications where outgassing might harm critical components such as electronic assemblies.

As usual, cost is a vital factor in choosing an adhesive, so that, except where the ultimate in strength, heat resistance, etc., is required, the vinyl phenolics would be favored since they are the cheapest structural adhesives. For structural bonding of metal parts, vinyl phenolics are required to meet Federal Specification MMM-A-132, Type I, Class 3. A qualified products list is available, showing which adhesives meet the specification.

NITRILE RUBBER PHENOLICS[7]

Nitrile rubber, also known as Buna N, was developed in Germany and became available in the United States about 1940. Adhesive

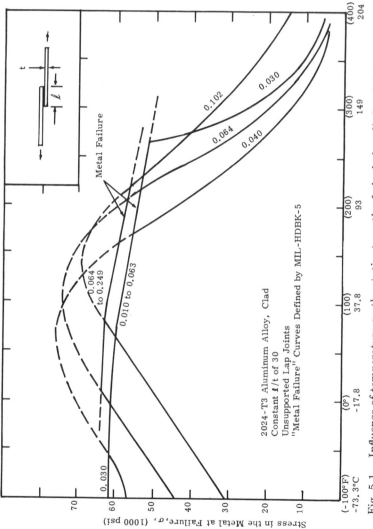

Fig. 5-1. Influence of temperature on the static strength of vinyl phenolic bonded lap joints.

formulations using nitrile rubber phenolic blends were developed in the early 1950's.[8]

Nitrile rubber, adhesive formulations make bonded structures more resistant to high temperatures and to oil. They are compatible with a wide range of resin modifiers and make it possible to bond successfully to highly-plasticized vinyl products. Their versatility widens the field of applications to include not only structural bonding in the aircraft and automobile industries, but also general bonding of nitrile rubber, polyvinyl chloride films, leather, wood, synthetics (both thermosetting and thermoplastic), and paper products.

As general purpose adhesives with good resistance to oil and petroleum products, nitrile rubber formulations qualify for MIL-A-5092, Type III.

Structural Nitriles

As structural adhesives with intermediate temperature resistance up to 177C (350F), nitrile rubber phenolics qualify for MMM-A-132, Type II. For structural-metal applications the high-strength phenolic, modified nitrile-rubber formulations are the most important. The effect of phenolic resin on nitrile rubber properties is shown in Table 5-2.

Table 5-2. Effect of Phenolic Resin on Nitrile-Rubber Properties*

Materials	Parts per Hundred					
"Hycar" OR-25EP (nitrile rubber)	100		100		100	
Phenolic resin	50		80		100	
Zinc oxide	5		5		5	
Sulfur	1.5		1.5		1.5	
Benzothiazol disulfide	1.5		1.5		1.5	
Stearic acid	1.5		1.5		1.5	
	Minutes cure at 154C (310F)					
	15	45	15	45	15	45
	psi					
Ultimate tensile-strength	2700	2900	3350	3800	4400	4250
Ultimate elongation	250	200	150	150	100	50

*Taken from Reference 9.

Increase in the phenolic-resin content increases strength and decreases elongation. Structural adhesive formulations are usually based on the high-strength end of the formulations in Table 5-2, that is, equal parts of nitrile rubber and phenolic resin. Even the lowest elongation of 50 percent is high compared to other structural adhesives. Thus, it can easily be seen why this adhesive has the lowest modulus of the structural adhesives listed in Table 4-1.

Table 5-3. Physical Properties of Nitrile-Phenolic/Metal-Bonds*

Property and Test Conditions	Specif Avg Requirement	Results			Control Avg
		Avg	High	Low	
Tensile shear, psi Normal temp 24C (75F)	2500	4400	5000	3800	...
Tensile shear, psi 10 min @ 82C (180F)	1250	2900	3325	2640	...
Tensile shear, psi 10 min @ 163C (325F)	None	1510	1840	1200	...
Tensile shear, psi 10 min. @ −55C (−67F)	2500	5150	5280	4960	...
Fatigue strength Normal temp 24C (75F)	600 psi 10^7 cycles	600 psi over 10^7 cycles			...
Creep-rupture 1600 psi @ normal temp. 24C (75F) (max.)	192 hrs 0.015 in. deformation	1600 psi 0.002 in. deformation			...
Creep-rupture 800 psi @ 82C (180F) (max.)	192 hrs 0.015 in. deformation	800 psi 0.0021 in. deformation			...
Tensile shear, psi Normal temp 24C (75F) after 30 days of salt-water spray	2250	4014	4840	2600	4370
Tensile shear, psi Normal temp 24C (75F) after 30 days at 49C (120F) 95–100% RH	2250	4160	4315	3980	4010
Tensile shear, psi Normal temp 24C (75F) after 7 days of immersion in JP-4 fuel (MIL-J-5624)	2250	4120	4260	3930	4035
After 7 days of immersion in anti-icing fluid (MIL-F-5566)	2250	4176	4305	4030	4080
After 7 days of immersion in hydraulic oil (MIL-H-5606)	2250	4142	4340	3820	4030
After 7 days of immersion in hydrocarbon (MIL-S-3136) (Type III)	2250	4275	4410	4175	4300
After 30 days of immersion in tap water	2250	4157	4270	4070	4050

* Tested in conformance with Fed. Specif. MMM-A-132, Type II, 0.064-inch, 2024-TC Alclad, bonded at ½-inch depth of lap. Cure: 60 minutes at 177C (350F), 40-psi pressure.

Nitrile-rubber phenolic has made and is still making a name for itself as one of the most durable and toughest materials so far developed by the adhesives industry. Its resistance to water has already been describ?d in Chapter 2 and illustrated in Fig. 2-2; it exhibited cohesive failure at 2700 psi even after two years of continuous exposure to high humidity and after two years of continuous immersion in water.

The overall performance of a nitrile rubber phenolic adhesive is shown in Table 5-3, where it is compared to Federal Specification MMM-A-132, Type II.

The data in Table 5-3 show good values—4,000 to 5,000 psi shear at ambient conditions with more than a 50 percent decrease at 163C (325F). Resistance to aqueous and organic solvents is shown by the shear values in excess of 4,000 psi after exposure. Toughness is illustrated in Table 5-4 by the medium-to-high values in peel, ex-

Table 5-4. Metal-to-Metal Peel on Nitrile Phenolic Adhesive

Test Method	Test Temp	Peel, lbs /in.	
		Avg	Range
Metal-to-Metal climbing-drum peel	24C (75F)	90	100 to 80
Metal-to-Metal climbing-drum peel	82C (180F)	73	76 to 72
Metal-to-Metal climbing-drum peel	121C (250F)	63	65 to 60
Metal-to-Metal climbing-drum peel	−55C (−67F)	22.5	30 to 18
T-peel, lbs/in.	24C (75F)	45	53 to 37
T-peel, lbs/in.	82C (180F)	25	29 to 22
T-peel, lbs/in.	121C (250F)	17	21 to 14
T-peel, lbs/in.	−55C (−67F)	4	5 to 4

cept for the value at −55C (−67F) which approaches the peel strength of an unmodified epoxy adhesive. Table 5-4 also emphasizes the difference in peel values obtained by the drum method and the T-peel method and points up the importance of defining the peel method when giving peel values for an adhesive. The drum peel and the Bell peel methods (see Table 4-3) will give higher values than the T-peel method.

Durability

Further evidence of the durability of the nitrile phenolic adhesive in terms of temperature resistance is given in Fig. 5-2 by Kuno.[10]

After 30 months at 120C (250F), the nitrile phenolic has a higher shear strength at this temperature than either the epoxy phenolic or the nylon epoxy. However, the nylon epoxy shows a remarkable steady increase in shear strength over the 30-month period.

At 177C (350F) the nitrile phenolic retains its superiority over the epoxy phenolic, while the nylon epoxy has dropped out of contention. However, both the nylon epoxy and epoxy phenolic do have outstanding properties in specific areas which will be described later on.

Processing

As to processing properties, the nitrile phenolics are intermediate between the vinyl phenolics and the neoprene phenolics. They are available as solutions in organic solvents and in film form. The solutions can be used alone as adhesives or as primers. (The primer function has already been described in Table 3-7, on surface preparation.) As an adhesive, the solution can be thinned out to any desired solids content for brushing or spraying. It is then air dried and oven dried per the manufacturer's recommendations. Proper preassembly drying is important to minimize the amount of retained volatile content; otherwise excessive outgassing during cure will seriously impair the bond strength.

The nitrile phenolics exhibit very little flow during cure. The flow can be regulated by the amount of pressure used. The pressure can be as low as 25 psi or greater than 200 psi. Pressure recommendations can be confusing in that, while 25 psi may be sufficient to cause flow and bonding, much more pressure may be needed to bring the mating surfaces together.

Because of their low flow, nitrile phenolics are used primarily for metal-to-metal and laminate bonding and not for metal-skin-to-honeycomb application. They would be especially useful where excess squeeze-out was objectionable or harmful.

A striking example of specialized adhesive application is found in the aluminum honeycomb panels of the General Dynamics plane, the B-58 (Hustler), where an epoxy phenolic adhesive was used to bond the aluminum honeycomb to the aluminum skin because of its

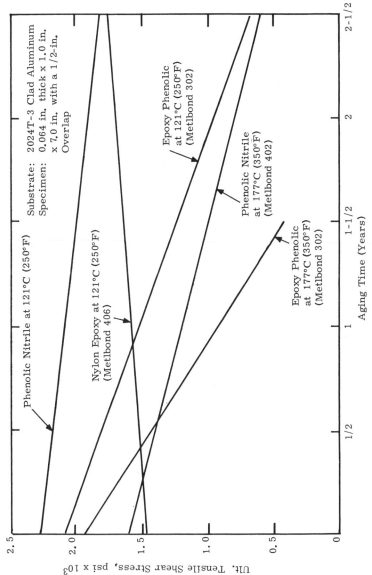

Fig. 5-2.   Effect of long-time aging on the different classes of structural adhesives.

low-creep properties at temperatures up to 260C (500F), while a nitrile phenolic adhesive was used for the metal-to-metal bond around the panel edge because of its high peel-strength coupled with high-temperature stability.[11]

EPOXY PHENOLICS

Epoxy phenolics, the last of the phenolic copolymers in general use as structural adhesives, were first developed about 1955.[12] They were designed to meet the need for a more temperature-resistant adhesive as aircraft surfaces were exposed to increased temperatures at supersonic speeds. The epoxy phenolics meet the requirements for Federal Specification MMM-A-132, Types III and IV, that is, exposure to temperatures from 149C (300F) to 260C (500F) for times up to 192 hours.

A representative formulation for an epoxy phenolic would contain an epoxy resin with a phenolic resin as the main ingredients, with a high loading of aluminum powder possibly acting as an oxygen "getter," and an amine curing agent.[13]

Cure Temperatures

A unique feature of this type of adhesive is that the strengths shown in Table 5-5 are obtained with relatively low-temperature cures. The adhesives already described and the high-temperature epoxies which will be described later, require cures of 149C (300F) to 177C (350F) to achieve strengths which decrease rapidly above

Table 5-5. Effect of Cure-time and Temperature on Lap-shear Strength of Epoxy Phenolic

Cure Conditions		Test Temperature	
		Ambient	900F
Temperature	Time, hrs	Lap Shear Strength, psi	
74C (165F)	24	580 (3)*	96 (5)
	48	1360 (4)	260 (5)
93C (200F)	24	2540 (5)	390 (5)
	48	2560 (5)	420 (5)

* Lap-shear specimens prepared with sandblasted stainless steel per ASTM D1002. Figures in parentheses are number of specimens tested.

177C (350F). The epoxy phenolics, however, can be cured at temperatures as low as 74C (165F) and 93C (200F)—given sufficient time—to give reasonably good strengths at 482C (900F). The data in Table 5-5 indicate that 24- and 48-hour cures at 74C (165F) are incomplete. Cures of the same length at 93C (200F) give values which compare favorably with higher-temperature cures. Evidence of this type of reactivity is given by the short shelf-life of the film or one-part paste at ambient temperatures. The shelf life without refrigeration is about one month. Unless there is a rapid turnover of the stock on hand, epoxy phenolic adhesives should be stored at temperatures lower than −18C (0F).

Applications

As mentioned in the section on chemistry, the epoxy phenolics have volatile by-products on curing. This results in a tendency to foam, so that on curing, the adhesive expands and acts as a space-filler. Epoxy phenolics are useful for bonding assemblies where tolerances are large and bond lines thick. The adhesive finds wide use in the bonding of metal-to-metal and metal-to-aluminum honeycomb, and is also used with glass-reinforced resin laminates. Its foaming properties make it ideal for splicing honeycomb cores. It has found wide use in missile and aircraft applications.

Epoxy phenolics are available in film form with a glass-cloth scrim support and as one-part pastes with varying solids content, depending on the application. The film form would be used where thickness control is required. The paste would be used where bond thickness may vary. The paste, which may include volatile solvents, would require preassembly-drying at room temperature or at slightly elevated temperature to minimize foaming during cure.

While the construction of commercial supersonic airliners requires adhesives which are stable at temperatures up to 260C (500F) for thousands of hours rather than hundreds, the epoxy phenolics have unique properties which make them valuable members of the adhesive family.

EPOXIES

History

Epoxy resin adhesives have an interesting past, in that the basic resins were discovered in fields not concerned primarily with adhesives.[14] In 1936, Pierre Castan, of Switzerland, developed the first epoxy resin resulting from the reaction of epichlorohydrin and bisphenol A. He was specifically interested in the manufacture of dentures and other casting applications. His discovery was eventually licensed to Ciba Corp., Ltd. In 1939, the United States entered the picture in the person of Dr. Greenlee, working for Devoe Raynolds. He discovered the same resins in a search for a better coating material.

With this start, epoxy resin production in the United States increased to about 110 million pounds in 1965. It is interesting to note that coatings, one of the original goals, exceeds other uses by a wide margin. In 1963, out of a total epoxy production of 88,835,000 pounds, 39,398,000 pounds were used as coatings, while epoxy resins used as adhesives were second, with a production of 11,640,000 pounds. Epoxies are attractive as adhesives for a number of reasons:

1. They exhibit a wide range in viscosity, from less than 1000 centipoises to high-melting-point solids. They cure as adhesives with very little weight-loss or shrinkage during cure.
2. A wide range of physical properties in the cured adhesive can be obtained, along with a wide range of curing temperatures from room temperature to 177C (350F). With both filled and unfilled adhesives, contact pressure is usually sufficient to insure a good bond.
3. The resin wets polar, or high surface-energy solids, well. This, combined with low shrinkage on cure, results in good bonds.
4. The cured resins are good electrical insulators and have good resistance to chemical and environmental degradation.
5. The properties can be modified by the use of fillers and flexibilizers to fulfill a wide range of requirements.

Resin Types

Today epoxy resins are almost infinite in their variety. Lee and Neville list 25 different types, all having one thing in common, that is, that each molecule contains two or more epoxy groups $\left(\begin{smallmatrix} & O & \\ & /\backslash & \\ -C & - & C- \end{smallmatrix} \right)$.[15] With changes in chain length, each type may then evolve into a series varying from low-viscosity liquids to high-melting-point solids. A simplified structure of an epoxy molecule is shown in Fig. 5-3, picturing only the terminal epoxy group.

Curing Agent Types

Reference 15 lists 68 curing agents including the more commonly used amines, primary and secondary amines, and anhydrides. Only a few of these find wide use in adhesive formulations. Some typical reactions appear in Fig. 5-3. The primary amine reaction involves

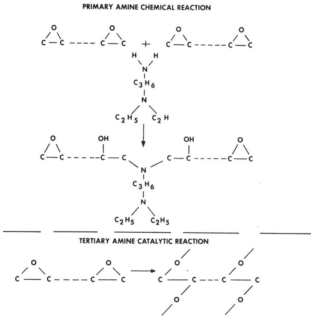

Fig. 5-3.   Epoxy resin polymerization chemical *vs.* catalytic.

a chemical exchange in which the amine hydrogen forms a hydroxyl group with the epoxy oxygen. This results in a carbon-nitrogen linkage. The tertiary amine behaves catalytically in breaking up the epoxy group so that the epoxies react with each other to form a series of ether linkages. Anhydride reactions (not shown here) behave somewhat like the catalytic reactions in that the oxygen-containing anhydride ring breaks up to react with the ruptured epoxy rings to form a network of ether and ester linkages. Tertiary amines are often used with anhydrides as catalytic agents to accelerate the reactions. As is evident in the reactions in Fig. 5-3, there are no volatile by-products, and thus, shrinkage due to polymerization is minimized.

The basic structure of the various epoxies may vary from low-molecular-weight structures which favor easy processing, to larger structures containing phenyl groups, as in the Novolac epoxies, which build in greater heat resistance.

As adhesives, epoxy resins are modified by the use of fillers, by curing agents which contain a flexibilizing capability, and by copolymerization with other resins.

Fillers

Fillers increase the modulus and decrease the coefficient of expansion. Fibrous fillers such as asbestos lend toughness to the cured adhesive. Table 5-6 gives electrical and modulus data for two filled and two unfilled epoxy formulations. While the two types have similar properties at room temperature with differences due to the fillers, at elevated temperatures the epoxy cured with diethanolamine softens readily with a corresponding decrease in modulus and volume resistivity. The epoxy with the aromatic-amine curing agent retains most of its properties, both physical and electrical, at the elevated temperatures. Diethanolamine-cured epoxy is usually cured at temperatures from 66C (150F) to 93C (200F), while the aromatic amine usually requires cures from 93C (200F) to 121C (250F).

Rigid, or nonflexibilized, epoxy adhesives have tensile-adhesion strength of about 6000 psi, lap-shear strength of about 3000 psi, and T-peel strength of less than 5 pli.[16] Rigid epoxies exhibit negligible creep. Because of the low peel-strength, joint designs using epoxy adhesives must minimize peel stresses.

Individual Characteristics

The great difference in strength characteristics with temperature for epoxy adhesives is shown in Fig. 5-4. Each curve reflects some change in epoxy resin, curing agent, and/or filler. It is interesting to note the difference between the curves representing epoxy adhesives and the characteristics of an epoxy phenolic adhesive (not shown). While the epoxies exhibit a wide variation in strength over a relatively narrow part of the temperature spectrum, the epoxy phenolic has an almost linear slope with considerably less variation

Table 5-6. Electrical and Physical Properties of Representative Epoxy Formulations

Formulation No.	Temperature		Temperature		
	Ambient	74C (165F)	Ambient	66C (151F)	149C (300F)
	Flexural Modulus, lbs/sq in./in.		Volume Resistivity, ohm/cm		
1	0.469×10^6	0.0485×10^6	6.7×10^{15}	7.7×10^{12}	6.4×10^6
2	1.21×10^6	0.0788×10^6	1.1×10^{15}	5.8×10^{12}	1.1×10^7
3	0.462×10^6	0.374×10^6	7.6×10^{15}	4.6×10^{15}	3.5×10^{11}
4	2.37×10^6	1.5×10^6 (estimated)	8.8×10^{15}	4.2×10^{15}	2.1×10^{12}

(1) Epon 828   100 parts by weight
Diethanolamine   12 parts by weight

(2) Epon 828   50 parts by weight
Mica   50 parts by weight
Diethanolamine   6 parts by weight

(3) Epon 828   100 parts by weight
Aromatic amine eutectic   20 parts by weight

(4) Epon 828   100 parts by weight
Alumina   300 parts by weight
Aromatic amine eutectic   20 parts by weight

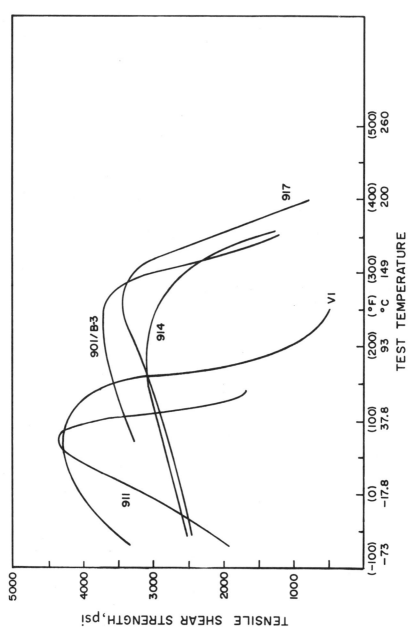

Fig. 5-4.   Tensile shear strength *vs.* test temperature for Epon adhesives. (*Courtesy of Shell Chemical Co., Adhesive Department.*)

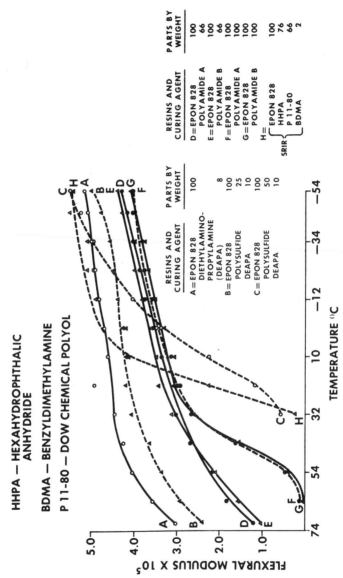

Fig. 5-5. Flexural modulus of flexibilized epoxy resins.

over a much greater temperature range. It has a lower, initial shear-strength (about 2500 psi) than any of the epoxies, but it retains a much greater percentage of it—about 2000 psi at 190C (375F)—as compared with the adhesives characterized in Fig. 5-4.

Since we seldom get something for nothing, it follows that high-temperature resistance usually requires higher-temperature cures. The high-temperature-cure adhesives offer several advantages; they have long-time stability or shelf life at room temperature; quite often they are one-part materials, which eliminates the need for weighing and mixing two or more components. This, in turn, reduces waste, since in weighing out and mixing the parts of a multi-component adhesive there is a tendency to make up more than is needed.

A widely used type of flexibilizing curing-agent is the polyamide. Because it can be used with good results in equal parts with some epoxy resins, it is commonly used in the twin-tube, epoxy-adhesive kits. This type is almost certainly present when the two tubes in the kit contain equal volumes of resin and hardener.

The effect of different curing agents on modulus is shown in Fig. 5-5, where flexural modulus is plotted against temperature. The polyamide curing agents which were evaluated show less change in properties with difference in concentration than do the polysulfide-modified epoxies.

A significant difference in behavior of the semirigid epoxies is that some, like the polysulfide-modified formulations, have a higher modulus at the low temperatures than the unmodified rigid formulations. Meanwhile, the polyamide-cured formulations retain some of their flexibility at the low temperatures.

Epoxy resins are compatible with a wide variety of other resins, so that many copolymer modifications are possible. Some of the oldest and best-known, the epoxy phenolics, have already been described. Epoxies have been combined with vinyls to increase flexibility. A copolymer which resulted in an adhesive with greatly increased peel-strength resulted from the combination of nylon and epoxy. This will be described later along with the nitrile rubber epoxies, which are probably the latest significant adhesives developed.

Applications

The development of epoxy resin adhesives with their contact-pressure assembly properties and with their low, intermediate, and high-temperature cures has given the engineer a much greater choice of materials for use as fasteners. Epoxies are used for bonding microelectronic parts, solar cells in satellites, and structural members and other parts of airplanes. Pipe railings can be assembled with epoxies.[17] Auto body repairs are made with epoxy resins. The abrasives industry bonds particle abrasives and abrasive bits. The jewels in watch works may be epoxy bonded. The alumina cutting tool shown in Fig. 5-6 was simplified by being bonded to a cold-rolled steel shaft. This not only saved money in terms of size of the alumina piece, but made the part much more resistant to impact blows. The adhesive used was a one-part epoxy paste using a dicyandiamide curing agent.

Fig. 5-6.   Alumina tip cutting tool bonded to steel shaft
with one-part paste epoxy adhesive.

The versatility of epoxy adhesives is illustrated by their use in optical assemblies. Here optical clarity and ability to transmit light with minimum absorption are the important properties. Epoxies with optical properties are available from the industry and a specification for such optical adhesives is in the process of being written by ASTM. When the applications require that optical or other glass assemblies be bonded to metal, highly flexibilized formulations are needed to make up for the difference in coefficient of expansion. Such formulations are available with elongations of over fifty percent using flexibilizers as described on page 106.

Thermoset molded parts and resin-glass laminates are replacing metal parts in many appliance applications and in lightweight structural parts where the high, strength-to-weight characteristics of laminates makes them preferable to metals. Epoxy adhesives are unequaled as fasteners for these materials. However, it must always be kept in mind that these materials usually have a mold-release finish which must be removed by some abrasive method in order to achieve a successful bond to the mold or laminate surface.

Machines have been developed for the automatic metering and mixing of adhesives and sealants, including epoxies. It is also possible to get epoxy adhesives premixed, deaerated and frozen in collapsible tubes of a size to fit any specific bonding operation.[18] Frozen adhesives make it possible to minimize the errors due to weighing and mixing for room-temperature and intermediate-temperature cure epoxies.

Epoxy Formulating Ingredients[19]

To a certain extent, epoxy resins lend themselves to "on the spot" formulating because they are relatively fluid; if they are too viscous, the viscosity can be decreased by warming to facilitate mixing-in fillers, flexibilizers, diluents, etc.

The most commonly-used epoxy resins have an epoxide equivalent of 195–205, meaning that each active epoxide group represents a molecular weight of 195–205. Each active hydrogen in a primary or secondary amine would react with one epoxy group. Thus, in calculating equivalent amounts of amine the molecular weight of the amine, in grams representing one active hydrogen, is combined with the molecular weight of epoxy resin in grams representing one epoxy group. The optimum concentration of tertiary amines (which are catalytic) must be determined empirically by evaluating the strength characteristics of a series of formulations where catalyst concentration is varied as in Fig. 5-7, where the optimum concentration of diethylamino propyl amine (DEAPA) is determined. DEAPA combines a tertiary amine with a primary amine, and the optimum concentration of about 10 parts per 100 of resin as indicated in Fig. 5-7, is considerably less than the calculated equivalent of 34 parts per 100 of resin. The equivalent concentration of an anhydride curing agent is calculated in the same manner, that is, the gram-molecular-weight per one epoxy group is added to the gram-molecular-weight per one anhydride group.

Some of the epoxy resins with epoxide equivalents of 195–205 and 180–220, available from various suppliers, are listed in Table 5-7.

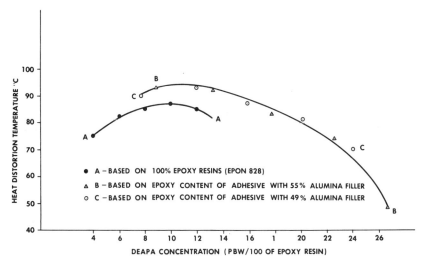

Fig. 5-7. Effect of curing agent concentration on deflection temperature.

Table 5-7. Epoxy Resins

Epoxide equivalent: 195 to 205. *Viscosity*: 12,000 to 16,000 centipoises

Commercial Designation	Source	Remarks
Epon 828	Shell Chemical Co.	All are relatively viscous; used where low outgassing and best high-temperature properties are required.
Hysol 2039	Hysol Corp.	
Araldite 6010	Ciba Products Co.	
DER 331	The Dow Chemical Co.	
Epotuf 6140	Reichhold Chemicals, Inc.	

Epoxide equivalent: 195 to 205. *Viscosity*: 500 to 800 centipoises.

Epon 815	Shell Chemical Co.	All are relatively fluid; used for impregnation applications for highly filled formulations and where thin glue lines are desired.
Hysol 2038	Hysol Corp.	
Araldite 506	Ciba Products Co.	
DER 334	The Dow Chemical Co.	
Epotuf 6130	Reichhold Chemicals, Inc.	

Novolac Epoxies—Epoxide equivalent: 180 to 220. *Viscosity*: semisolid

DEN 438	The Dow Chemical Co.	These are usually used for heat-resistant formulations.
Epon 1031	Shell Chemical Co.	

Curing Agents

Curing agents are of various types: Room-temperature curing agents for general purpose and quick-cure applications; intermediate-temperature curing agents for most applications; high-temperature curing agents where B-stage cures and high-temperature strengths are required. Some of the available curing agents (Concentration is given as the parts by weight per 100 grams of an epoxy with an epoxide equivalent of 195–205) are listed in Table 5-8.

Flexibilizers

Flexibilizers are sometimes added to adhesives in varying proportions to give a more resilient compound, for example, to epoxy sealers for use in expansion joints. They should be used with caution since they generally degrade electrical and physical properties. Concentration of the flexibilizer can be varied depending on flexibility desired, but it may also affect the concentration of the curing agent. Some available flexibilizers are: LP 3 from Thiokol Chemical Corp., Epon 871 & 872 from Shell Chemical, and DER 731 from the Dow Chemical Company.

Fillers

Fillers are important in reducing stress in the bond area by reducing the coefficient of expansion. In this respect it is the volume percent of filler which is important, not the weight percent. The percent of filler by weight is usually reported because it is simpler to measure by weight. Thus, in any basic approach to the problem of reducing coefficient of expansion one must think in terms of volume percent of filler along with density and coefficient of expansion of the filler. Fibrous fillers such as asbestos or glass fibers are very commonly used to lend toughness to an adhesive. For instance, practically all adhesives which qualify for MIL-A-8623 have asbestos or glass-fiber fillers.

All the materials listed in Table 5-9 (and many more), can be combined in different proportions to give adhesives varying from very flexible to very rigid. Some fillers, such as Bentone clay, result in thixotropic adhesives which can be applied on vertical or overhead surfaces. Others result in heavy self-leveling properties. Electrical, physical, and processing properties can be varied to suit a wide variety of applications using the additives mentioned above.

Table 5-8. Curing Agents for Epoxy Resins

Commercial or Chemical Designation	Source	Concentration, gms	Remarks
Diethylene triamine (DETA) Triethylene tetramine (TETA) Diethylene triamine adduct	Chemical Suppliers Epoxy Suppliers Chemical Suppliers Epoxy Suppliers Epoxy Suppliers	10 10 to 12 20 to 30	All have quick cures, i.e., 2 hrs @ 74C (165F) or 16 hrs @ room temperature. Not recommended for use near electrical contacts because of corrosion possibilities.
Polyamide	Epoxy Suppliers	40 to 100	Cures 2 hrs @ 74C (165F) or 16 hrs at room temperature. This type is used successfully over a wide concentration range giving semirigid cures in higher concentrations. Often used in twin-tube kits.
Diethylaminepropylamine (DEAPA)	Chemical Suppliers Epoxy Suppliers	6 to 8	One of the most widely used curing agents, especially for adhesives called out in MIL-A-8623, Type II. Cures in 4 hrs @ 74C (165F).
Curing Agent Z (aromatic-amine eutectic)	Shell Chemical Co.	20	Intermediate cure, 5 hrs @ 107C (225F). Good temperature resistance up to 149C (300F).
Metaphenylene diamine Methylene dianiline	Chemical Suppliers Epoxy Suppliers Chemical Suppliers Epoxy Suppliers	14 28	Characteristics similar to above. These are solids which must be melted before adding to epoxy.
Dicyandiamide	American Cyanamid Co.	4	Good shelf-life as curing agent in one-part pastes.
Methyl-nadic anhydride (MNA)	Chemical Suppliers Epoxy Suppliers Rohm & Haas Co.	80 to 90 plus 0.5 DMP-30	Liquid anhydride suitable for vacuum impregnation because of fluidity. Cure is 24 hrs @ 121C (250F).

Table 5-9. Additive Fillers for Adhesives

Filler	Source	Percent by Weight
Asbestos	Raybestos Manhattan, Inc.	About 30
Aluminum silicate	Edgar Brothers Co.	
Glass fiber	Corning Glass Co.	15 to 25
Silica gel (Cab-O-Sil®)	Cabot Corp.	5 to 20
Alumina	Aluminum Co. of America	66 to 110
Bentone clay	Baroid Division, National Lead Co.	10 to 20
Silica	United Clay Mines Corp.	50
Ferric oxide		66 to 110
Graphite	Acheson Colloid Co.	50
Acetylene black	Columbia Carbon Co.	10

Processing

Epoxy adhesives can be formulated to meet practically any processing requirements. They are available as low-viscosity, unfilled liquids in two parts for both room temperature and intermediate temperature cures, and one-part high temperature—135C to 177C (275F to 350F)—curing liquids. These would be used where thin glue lines are necessary or in close-fitting cylindrical assemblies. Solvents are sometimes added to facilitate spraying, although undiluted epoxy resins have been used with hot-spray equipment. When the parts to be bonded are warmed, fluid epoxy adhesives can be introduced into the bond by capillary action, thereby taking advantage of their excellent wetting properties.

Epoxy adhesives formulated to meet more rugged structural requirements are usually filled and are available as thixotropic pastes which require only contact pressure during assembly to obtain close fitting of the mating surfaces. They are good fillers and work well with coarse-fitting parts. These adhesives at their best are formulated to meet the requirements of specifications MIL-A-8623, Type III, Federal Specification MMM-A-132, Type I, Class 3, and Type II.

MIL-A-8623 is divided into three types, all of which have to test at 2500 psi, in shear at ambient. Requirements vary for other conditions depending on type. The specification requires that: Type I (two parts) shall cure at 20C-30C (68F-86F) or at 71C (160F) for one hour. Type II (two parts) shall cure at 31C to 99C (87F to 210F) for 2 hours. Type III (one part) shall cure above 99C (210F). Federal Specification MMM-A-132, superseding MIL-A-5090D, has similar performance requirements. No peel requirements are listed for this type of adhesive. Both specifications have qualified products lists (QPL) which list qualified adhesives and manufacturers.

TWO-PHASE, NITRILE RUBBER, PHENOLIC/EPOXY ADHESIVE

This type was developed as a film which is all epoxy adhesive on one surface and all nitrile-rubber phenolic on the other surface. As such, it overcomes some of the limitations of each by combining the toughness and high peel-strength of the nitrile phenolic with the good filleting characteristics of the epoxy. It is used to bond both metal-to-metal edge parts and skin-to-honeycomb sections. Care must be taken to lay up the assembly so that the epoxy surface is on the honeycomb side, while the nitrile phenolic is on the aluminum skin side. This type of adhesive has the same temperature limits as the nitrile phenolic and has peel strengths intermediate between epoxy and nitrile phenolic types with metal-to-metal, T-peel strengths of about 15 pli. It simplifies assembly practice by allowing the use of one type of adhesive for an entire sandwich panel.

NYLON EPOXIES

Probably the most significant offshoot of the epoxy adhesive family is the nylon epoxy or polyamide epoxy. This type is not to be confused with the two-part polyamide, room-temperature-cured epoxies which have already been described as semirigid, epoxy adhesives. The nylon epoxy is a blend of an epoxy resin with a soluble grade of nylon[20] and a high-temperature curing-agent.

The nylon epoxy adhesive development was a giant step forward in terms of strength increase, durability at temperatures up to 121C (250F), and filleting characteristics with aluminum honeycomb. The other adhesives described up to this point have all had shortcomings in the various areas of adhesive application. Table 5-7 gives representative values for the structural adhesives to illustrate the superiority of the nylon epoxies.

An overall survey of the values in Table 5-10 reveals that temperature resistance is the only area in which the nylon epoxies fall short of complete superiority. Yet even in this property, referring to Fig. 5-2, we see that the nylon epoxy adhesive, after two and one-half years at 121C (250F), is almost on a par with the nitrile-rubber phenolic. The nylon epoxy is sensitive to water, as we saw in Chapter 2; however, when it is exposed to the 30-day water immersion specified in MMM-A-132, shear strengths are still respectable with values of about 4000 psi.

Peel strength, both T-peel and sandwich, is where this material really stands out, with values of 100 pli and 250 lbs/3-inch width. Its ability to soften and wet so as to form a distinctive fillet on the aluminum honeycomb edge tends to cause aluminum failure in the lighter-gauge aluminum foil.

Figure 5-8 taken from Reference 10 shows that its excellent shear properties exist over a 316C (600F) temperature range from −240C (−400F) to +93C (+200F).

Applications

The nylon epoxies are used primarily in making structural aluminum honeycomb panels for aircraft. Their ability to bond well to the edge metal-to-metal sections, and to the metal-skin-honeycomb sections simplifies construction. Lower film weights are possible with this material, and its great strength in both peel and shear provides a greater safety factor. It is also useful in metal-to-metal bonds where high peel and shock resistance are required.

Nylon epoxies are available in unsupported film thicknesses from 0.001- to 0.010-inch. Tack primers have been developed which simplify assembly under ambient conditions and at elevated temperatures. Elevated-temperature tack primers were developed for complex assemblies such as helicopter rotor blades. These are primed and assembled with the adhesive film in place. Electrically-heated clamps are used to activate the tack primer which then holds the assembly firmly together during the bagging operation and until heat and pressure are applied in the autoclave.

The usual cure conditions for nylon epoxy films are a temperature rise time to 177C (350F) of 60 minutes or less and a dwell of 60 minutes at 177C (350F). Since the adhesive film softens considerably before cure, only enough pressure is needed to insure contact of the mating surfaces.

The temperature-rise time is critical when bonding to honeycomb. If the temperature-rise time is too long, the film will cure

Table 5-10. Adhesives Types and Properties

Adhesive Spec. Type	Maximum Operational Temp.	Strength				Cure Pressure, psi	Cure Temperature
		Tensile Lap-shear, psi	Tensile, psi	T-Peel, pli	Sandwich Peel, lbs/in.		
Nylon epoxy MMM-A-132 Type I, Class 1	100 to 125C (212 to 257F)	6000	8000	75 to 100	200	10	177C (350F)
Nitrile-rubber phenolic MMM-A-132 Type II	175C (347F)	4000	3000	35	Not recommended	50 to 100	149C (300F)
Vinyl phenolic MMM-A-132 Type I, Class 2	100C (212F)	3500	4000	30	60	250	177C (350F)
Epoxy per MIL-A-8623 Type II	100C (212F)	3500	6000	<5	...	Contact	74C (165F)
Epoxy phenolic MMM-A-132 Type III and IV	260C (500F)	3000	2500	<5	30	25	149C (300F)
Nitrile-rubber epoxy	125C (257F)	5000	...	25	70	25	94C (200F)
Polyimide	316C (600F)	2000	...	<10	...	40 to 100	260C (500 to 700F)

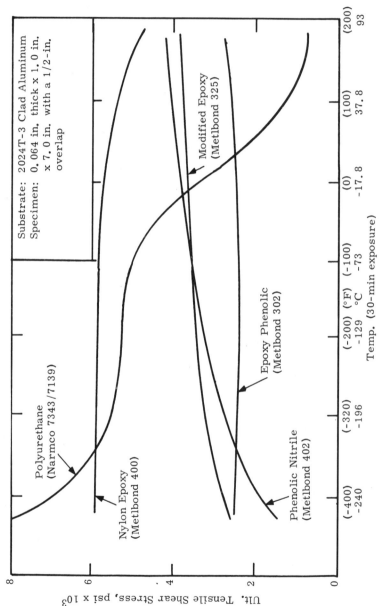

Fig. 5-8. Effect of cryogenic temperatures on different classes of structural adhesives.

before softening and poor filleting will result. In metal-to-metal bonds temperature-rise time is not as critical.

Cure parameters for each adhesive type are usually carefully evaluated by manufacturers of structural adhesives, and consumers should take full advantage of all the data available, since this could prevent quite a few production headaches.

NITRILE-RUBBER EPOXY ADHESIVES

This type is one of the latest developments in structural adhesives. The film form first saw the light of day early in 1967; the paste form was available somewhat earlier.

While properties will vary depending on formulation variables, this material will fill an important gap in the structural-adhesives spectrum in that it can be cured at temperatures as low as 82C (180F) and pressures as low as 10 psi. However, the lower-temperature cure (82C) does require a longer time, as shown in Table 5-11,

Table 5-11. Shear Strength as a Function of Cure Cycle at 82C (180F) for Nitrile-Rubber Epoxy Adhesive

Panel Number	Cure cycle	Temperature		
		RT	82C (180F)	121C (250F)
		Shear Strength, psi		
123–7118 123–7119	60 min to 82C (180F) 4 hrs at 82C (180F) 40 psi	1350 1300 1000 1450 1475 1235	Broke in jaws Broke in jaws Broke in jaws Broke in jaws Broke in jaws Broke in jaws	Broke in jaws Broke in jaws Broke in jaws Broke in jaws Broke in jaws Broke in jaws
123–7116 123–7117	60 min to 82C (180F) 6 hrs at 82C (180F) 40 psi	5100 5050 5100 4720 4600 4700	2255 2260 1200 1360	1030 1000 600 750 . . . . . .
123–7114 123–7115	60 min to 82C (180F) 8 hrs at 82C (180F) 40 psi	5360 5360 4800 5300 5200 5300	3190 3150 3000 2760	1550 900 1580 1580 . . . . . .

Material: 0.064-inch 2024-T3 Alclad—Lap-Shear Specimen per ASTM D-1002
Data Supplied by Bloomingdale Dept., American Cyanamid Co.

where 8 hours at 82C (180F) are seen to result in strengths comparable to those resulting from 250F cures. A glance at Table 5-10 shows that it is intermediate in peel-strength with a T-peel value of 25 pli, good in shear at 5000 psi, with a maximum operating temperature of 125C (257F).

The only penalty involved in using this adhesive is that it requires refrigerated storage. Recommended storage temperature is −17.8C (0F) or less. Nitrile-rubber epoxies, like the nylon epoxies, fillet well when bonded to honeycomb and thus can be used for both the edge bonding and honeycomb in structural lightweight-panel construction.

SUPERSONIC ADHESIVES

Until the advent of the Mach 3 (2100 mph) supersonic airplane, 177C (350F) was the highest operational temperature deemed necessary for existing aircraft. The epoxy phenolics, which could take 260C (500F) for relatively short times (hundreds of hours), were considered good, high-temperature adhesives.

However, Mach 3 supersonic transports have generated a need for materials which will withstand skin temperatures from 316C (600F) to 204C (400F) for at least 30,000 hours.[21] These requirements are testing to the ultimate the capability of presently available, nonmetallic materials.[22] Stainless steel and titanium construction with brazed honeycomb panels will undoubtedly bear the brunt of these high-temperature requirements, but the weight saving involved in using adhesives provides an incentive for the development of synthetic resin materials for similar use.

The excellent performance of the nitrile rubber phenolic adhesives at 177C (350F) for two and one-half years, as shown in Fig. 5-2, will undoubtedly suggest use of these materials away from the extreme skin conditions.

The present outstanding candidate for the high temperature—260C to 316C, (500F to 600F)—structural adhesive to be used in supersonic aircraft construction is the polyimide type. Quite a few polyimides are available, and both government and aircraft laboratories are hard at work optimizing the various formulations and screening out the better ones.

Results of test work with one of the better polyimide adhesives are shown in Figs. 5-9 and 5-10.[23] These show the results of aging lap-shear specimens, made with stainless steel, at elevated tempera-

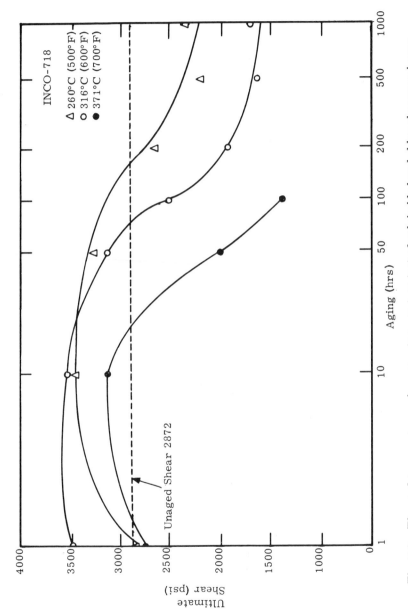

Fig. 5-9.   Elevated temperature aging room temperature test of polyimide bonded lap shear specimens.

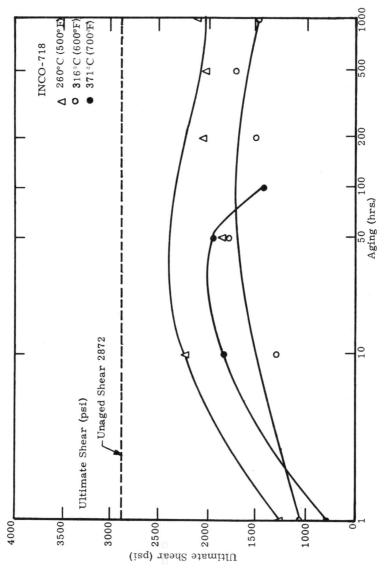

Fig. 5-10. Elevated temperature aging and tests of polyimide bonded lap shear specimens.

tures, 260C, 316C, and 371C (500F, 600F, and 700F) and testing at room temperature (Fig. 5-9) and at the aging temperature (Fig. 5-10). The data as plotted show that 371C (700F) is definitely too high a temperature for extended exposure. At 260C (500F) and at 316C (600F) the lap-shear values at room temperature and at the aging temperature may be leveling off at a respectable 500 psi, or better.

The polyimide adhesives are available as film, as pastes, and as unformulated resins. Processing temperatures and pressures increase considerably. Cure temperatures range from 260C (500F) to 371C (700F) at pressures of 40 to 200 psi.

Initial cost and processing expense are high for this type. In time both quality processing conditions and cost should improve to make this material more attractive.

MATERIAL COST

The price of adhesives is dependent on development costs and volume requirement. Adhesives in film form require more development and are more expensive than paste or liquid types. Film adhesives are sold by the square foot, paste adhesives by the pound.

Type	Price (per sq ft)
Polyimide	$4.00 to 5.00
Nitrile epoxy	0.90 to 0.95
Nitrile phenolic	0.85 to 0.95
Nylon epoxy	0.80 to 0.90
Epoxy phenolic	0.60 to 0.70
Modified epoxy	0.50 to 0.60
Vinyl phenolic	0.15 to 0.20

Bulk adhesives would be about $1.25 to $3.00 per lb in 5-gallon lots depending on type and solids content. Since many are sold as dilute solutions, this figure may vary considerably.

Some of the manufacturers of structural adhesives are listed below:

Adhesive Engineering, San Carlos, California
Aero Research Limited, Cambridge-Duxford, England
American Cyanamid Co., Bloomingdale Department, Havre de Grace, Maryland
Armstrong Cork Co., Lancaster, Pennsylvania

Armstrong Products Co., Warsaw, Indiana

Cycleweld Cement Products Div., Chrysler Corp., Trenton, Michigan

Emerson and Cuming, Inc., Canton, Massachusetts

Epoxylite Corp., El Monte, California

Furane Plastics Inc., Los Angeles, California

B. F. Goodrich Company, Akron 18, Ohio

Houghton Laboratories, Olean, New York

Minnesota Mining and Mfg. Co., Detroit, Michigan

Narmco Resins and Coatings Co., Costa Mesa, California

Pittsburgh Plate Glass Co., Bloomfield, New Jersey

Shell Chemical Corporation, Adhesives Div., Pittsburg, California

Further information on adhesives suppliers can be found in the following publications:

1968 Adhesives Redbook, New York: Palmerton Publishing Company

Handbook of Adhesives, New York: Reinhold Publishing Corporation

REFERENCES

1. Martin, R. W. *The Chemistry of Phenolic Resins.* New York: John Wiley & Sons, Inc., 1956.
2. Lavin, E., and Snelgrove, J. A. "Polyvinyl Acetal Adhesives." In *Handbook of Adhesives*, edited by I. Skeist. New York: Reinhold Publishing Corp., 1962.
3. Delmonte, J. *The Technology of Adhesives.* New York: Reinhold Publishing Corp., 1947.
4. Bake, L. S. "Neoprene Cements." In *Handbook of Adhesives*, edited by I. Skeist. New York: Reinhold Publishing Corp., 1962.
5. Available from: Bloomingdale Dept., American Cyanamid Co.
6. DeLollis, N. J. "Structural Adhesives — Characteristics and Application." In *Symposium on Adhesives for Structural Applications.* New York: Interscience Publishers, 1962.
7. Brown, H. P., and Anderson, J. F. "Nitrile Rubber Adhesives." In *Handbook of Adhesives*, edited by I. Skeist. New York: Reinhold Publishing Corp., 1962.
8. Engel, H. C. *Improved Structural Adhesives for Bonding Metals.* WADC Tech. Rept. 52–156, June 1952.
9. B. F. Goodrich Chemical Company. *Hycar Phenolic Blends.* Service Bulletin H-4, September 1950.
10. Kuno, J. K. "Comparison of Adhesive Classes for Structural Bonding at Ultrahigh and Cryogenic Temperature Extremes." Paper read at Society of Aerospace Materials and Process Engineers, Materials Symposium, Los Angeles, California, May 20–22, 1964.

11. Bandaruk, W. "B-58 Uses Bonded Honeycomb in Primary Structures," *Aviation Age*, 28 (3) 72, September 1957.
12. Black, J. M., and Blomquist, R. F., "Metal-Bonding Adhesives for High Temperature Service." *Modern Plastics.* 33 (10) 225, June 1956.
13. DeLollis, N. J. "Structural Metal Bonding." In *Handbook of Adhesives*, edited by I. Skeist. New York: Reinhold Publishing Corp., 1962.
14. Lee, H., and Neville, K. *Handbook of Epoxy Resins.* New York: McGraw-Hill Book Company, Inc., 1967.
15. Lee, H., and Neville, K. "New Developments in Epoxy Resins," *Insulation*, (in 6 parts), December 1960 to May 1961.
16. DeLollis, N. J. "Adhesives and Sealants in Electronics." *IEEE Transactions*, Vol. PMP-1 No. 3 (1965): 4–16.
17. "Epoxy Bonds Railing Joints." *Maintenance*, October 1964.
18. Ablestik Adhesive Co., Gardena, California.
19. Skeist, I. "Epoxy Resin Adhesives." In *Handbook of Adhesives*, edited by I. Skeist, New York: Reinhold Publishing Corp., 1962.
20. "Strengthening Adhesives with Nylon." *Dupont Magazine*, Vol. 57, No. 1 (1963): 12–15.
21. Raring, R. H. "Materials for Wings and Fuselage." *Materials Research and Standards*, October 1963, pp. 810–814.
22. Hightchew, H. E. "Nonmetallic Materials." *Materials Research and Standards*, October 1963, pp. 815–819.
23. Mahoney, J. W. *Optimization and Evaluation of High Temperature Structural Adhesives*, Technical Report AFML-TR-66-198, Wright-Patterson AF Base, Ohio: Air Force Materials Laboratory, September 1966.

CHAPTER 6

Miscellaneous Adhesive Materials

The adhesive materials already described are designated as metal structural adhesives because, properly used, they can stress metals beyond their elastic limit. However, a large number of adhesives have been developed which can be used with metals but which emphasize convenience in use and application rather than strength. These are primarily one-part, air-dry or room-temperature curing types. One exception is the epoxy twin-tube kit, which is a two-part epoxy that has been simplified for home and general use by packaging the two parts in collapsible tubes. The orifice sizes of the two tubes have been designed so that the user will squeeze out equal lengths for mixing prior to application. This eliminates the need for balances and weighing of exact amounts.

The general use types of adhesives are listed below with some comments on strength properties and limitations.

These twin-tube epoxy kits are probably the only two-part adhesive types retailed for home consumption. They are convenient to use, and the collapsible-tube packaging is made possible because the resin/hardener concentration can be varied between wide limits and still achieve useful cures. The curing agents are usually polyamide types resulting in rigid to semirigid cures. These bond well to a wide variety of materials including metals, ceramics, glass, and wood. They should be used primarily for high-strength bonds to rigid materials in joints which will not be exposed to peel or cleavage stresses. They will cure equally well in thin or thick bonds, being good filler-cements because of their low volatile content. The more rigid the application requirements, the more important it becomes to control closely the amounts measured out and mixed. This type is not usually heat resistant and tends to soften at relatively low temperatures so that it should not be used much above 66C (150F).

Not all twin-tube-kit epoxies are the same, as is shown in Table 6-1, which gives strength values for some of these epoxies after various conditions of exposure and at three different temperatures. Kit D, formulated to have a very quick cure, never developed very high strength. Kit C was only slightly better. The other five were all somewhat better, but still relatively weak at 74C, thereby emphasizing the precaution that this type should not be used at elevated temperatures. Epoxy Kit E, cured well enough after five days at room temperature, to reach a shear strength of 3100 psi. The best guide where strength at ambient temperatures is required is to select twin-tube kits which are packaged in equal-volume tubes.

ALKYL CYANOACRYLATES[1, 2]

This was one of the most highly publicized adhesives at the time of its development because of its ability to cure within minutes after application and assembly. Strength development is so rapid that within a few hours a tensile adhesion strength of more than 2000 psi may result.[1] An alkaline surface accelerates the cure; an acid surface may slow it down. Table 6-2 taken from Reference 3 gives some idea of rate-of-strength development under favorable conditions.

The material is formulated with an inhibitor so that it is stable in bulk packaging and does not cure in the presence of air. Since its cure does depend on surface catalysis, its major defect is that in bond thicknesses of 0.002-inch or greater, it cures very slowly if at all. Some surfaces, such as those plated with cadmium and nickel, tend to be passive so that cures are slow. These surfaces may be activated by coating with a one-percent solution in acetone of an amine such as methyldiethanolamine. While slightly alkaline surfaces cure the bond rapidly, water—especially alkaline water—tends to degrade the cured bond. The adhesive performs well at temperatures up to 100C (212F). A solvent for the cured adhesive is N, N, dimethylformamide. The adhesive which is squeezed out and exposed to air at the edge of a bond does not cure and is volatile. It is also an eye irritant, so it should be used in a well-ventilated area.

Disadvantages

As was demonstrated in the publicity accompanying the introduction of this adhesive, it has ideal basic properties: it is a one-

Table 6-1. Shear Strengths[a] of Twin-Tube Kit Epoxy Adhesives

Adhesive	Strength at Various Temperatures after Minimum 5-day Cure				Temp Cycling 6 Cycles	
	Test Temperature of Controls[b]				Strength at Room Temp after Exposure to:	
	−54C (−65F)	Room Temp	74C (165F)	Humidity[c]	−54C (−65F) to +74C (165F)	2 hrs @ 74C (165F)
	pounds per sq in.				pounds per sq in.	
A	875	1197	434	2420	2992	1572
B	993	1451	158	1270	2694	2764
C	670	607	363	593	2451	988
D (10-minute pot life)	342	582	295	925	870	858
E	1221	3100	270	1446	3100	2687
F	1082	1732	339	1030	2655	1499
G	950	1981	727	2178	2654	2006

[a] Aluminum tensile shear-specimens per ASTM D-1002.
[b] Minimum 5-day cure before testing.
[c] Humidity cycle consists of 20-day exposure to minimum 93 percent RH with temperature cycling from 32C (89F) to 65C (149F) every 48 hours.

Table 6-2. Tensile Properties of Metal-to-Metal Bonds for
Cyanoacrylate Adhesive*

Type of Metal Bond	Cure Time at Room Temperature							
	2 hrs	24 hrs	48 hrs	72 hrs	2 weeks	4 weeks	9 weeks	6 months
Steel-to-Steel Bonds:								
Tensile strength, psi[a]	2005	...	5030	...	...	4030	5180	6500
Shear strength, psi[b]	...	1735	2200	...	...	...	...	...
Aluminum to-Aluminum Bonds:								
Shear strength, psi[c]	...	2400	2600	2700	2700	...	...	...

*Taken from Reference 3
[a]ASTM D-897-49
[b]Modification of ASTM D-1002-49T
[c]Modification of MIL-A-5090B

part formulation with long shelf-life in its container, and it cures almost immediately once it is applied and the assembly is brought together. However, as has been said before, seldom do we get something for nothing, and this adhesive has certain drawbacks: surface passivity impedes curing; the material cures slowly in thick glue lines, is volatile, and is sensitive to moisture and alkalinity. Therefore, it must be used with caution and only after full evaluation for the intended application.

This adhesive probably finds its greatest use where production speed is an overriding requirement, especially in small parts including metal bonds to metals and to plastics. One should be very cautious with the use of cyanoacrylates in critical bonds and bonded assemblies where outgassing may cause problems. A thorough evaluation should precede any extensive use. Table 6-3 shows its sensitivity to bond thickness.

Table 6-3. Shear-Strength of Bonds made with Cyanoacrylate
Adhesive between Anodized Dural Surfaces

Adhesive	Glue Gap	Shear Strength After 2-hr Cure	Shear Strength After 24-hr Cure
	inches	psi	psi
Adhesive Alone	0.0005	63	1258
	0.002	81	748
	0.005	60	350
	0.007	68	545
	0.010	37	453
Cyanoacrylate (25 parts) / Cab-O-Sil® (1 part), Supplied by Cabot Corp.	0.0005	234	1143
	0.002	96	892
	0.005	48	519
	0.007	30	260
	0.010	47	372

A remarkable property of this adhesive is that it cures instantly on contact with human skin and tissues. This property has resulted in considerable research directed toward using it to replace sutures in operations on the human body.

POLYSULFIDES

Two-part polysulfides will be described in detail in Chapter 7. One-part polysulfides were developed to make polysulfides more competitive in terms of convenience of use in the building trades. The one-part formulation retains the good-aging elastomeric properties typical of polysulfides and eliminates the need for weighing and mixing. As a caulking sealant it bonds well to the various materials used in modern construction. It was recently recommended for sealing the joints of the many sections of an outdoor, galvanized-steel water conduit. It was considered capable of maintaining the seal through the daily and seasonal variations in temperature while allowing the joints to move as the metal sections expanded and contracted.

SILICONES

Two-part silicones will be described in Chapter 7. One-part silicones have become so prevalent that they have invaded the household as well as the factory. One-part room-temperature silicones are made possible through the use of blocking agents which inactivate the curing agent until atmospheric moisture hydrolyzes the chemically-blocked system thereby freeing the curing agent and allowing the inhibiting groups to depart as a volatile by-product. Because of the resulting polar groups, the one-part silicones have so-called "natural" adhesion and quite often can be used without a primer.

These materials perform well as adhesives for bonding silicone rubber to itself and to nonsilicone surfaces. However, unless they can be exposed to humidity for a long enough time to initiate the absorption of water and hydrolysis of the curing agent, it is not advisable to use these to bond impermeable materials to each other. The primary application is as an edge sealant in high temperatures, up to 250C (482F), and environment resistant joints in the aircraft, automobile, and construction industries. These adhesives are beginning to find wide use in home caulking jobs and even as boat caulking material.

RUBBER-BASE SOLVENT TYPES

These are probably the most widely used adhesives, forming good bonds with most organic, inorganic, and metallic adherends. The neoprene and nitrile types have already been described in the chapter on structural adhesives. They have also been described as meeting the requirements for MIL-A-5092 Types II and III, respectively. Natural-rubber adhesive formulations would qualify as Type I in MIL-A-5092. This type would qualify for general use in the least critical areas of bonding applications since natural rubber has less resistance to atmospheric exposure than neoprene and less resistance to petroleum fuels and solvents in general, than either neoprene or nitrile rubber.

Reclaimed rubber is available in solutions varying from syrupy liquids with 30-percent solids content to heavy-bodied mastics of 80-percent solids content. This rubber is reclaimed mostly from old tires and has a high percentage of butadiene styrene. The mastic form would find wide application where poorly matching or rough surfaces make it necessary to use a good, filler type of adhesive.

ANAEROBIC POLYESTERS

Like the cyanoacrylates already described in this section, this type does not cure in the presence of air and is catalyzed by the surfaces to which it bonds. The principal use is as a staking compound in threaded assemblies, and it is around this application that a specification has been written—MIL-S-22473, which is unique in that it lists the various grades by their proprietary designation. (This is possible because there is only one supplier for this material.[4])

From an initial four or five different grades this material is now available in fourteen grades varying in viscosity from 10 cps to 10,000 cps and varying in torque strength from 150 to 375 inch-pounds, through 10 to 25 inch-pounds. It can be used to bond cylindrical bearings in place and as a conventional adhesive to join close-fitting metal surfaces.

Since surfaces plated with metals such as cadmium or nickel tend to be noncatalytic, activators have been developed which, when applied as a thin coating and air-dried, will accelerate the

Table 6-4. Physical Properties of Various Grades of Thread Sealants

Grade	Locking Torque (inch-pounds)	Viscosity (centipoises)	Color Code (see 3.3.1)*
AA	150/375	10 to 25	Green
A	100/250	10 to 25	Red
D	100/250	40 to 80	Orange
AV	100/250	100 to 250	Red
AVV	100/250	1,000 to 10,000	Red
B	70/175	100 to 200	Yellow
C	40/100	10 to 25	Blue
CV	40/100	100 to 250	Blue
CVV	40/100	1,000 to 10,000	Blue
E	20/50	10 to 25	Purple
EV	20/50	100 to 250	Purple
H	10/25	10 to 25	Brown
HV	10/25	100 to 250	Brown
HVV	10/25	1,000 to 10,000	Brown

* Taken from MIL-S-22473.

cure considerably. Table 6-4 taken from MIL-S-22473 illustrates the strengths developed with these materials in terms of holding torque in threaded joints.

VINYL PLASTISOLS[5]

Plastisols are not usually considered to have adhesive properties, but properly formulated they are very tough, abrasion-resistant materials. They have found use in the automobile industry where their ability to bond to oily steel surfaces makes them unique. This is an instance where an adhesive was specifically designed to fit into production-line assembly and simplify assembly-line bonding by eliminating the usual elaborate cleaning prior to bonding.

REFERENCES

1. Coover, H. W., Jr. "Cyanoacrylate Adhesives." In *Handbook of Adhesives*, edited by I. Skeist. New York: Reinhold Publishing Corp., 1962.
2. Supplied by Eastman Kodak Company as Eastman 910 Adhesive, Marketed by Armstrong Cork Company, Industrial Division.
3. "Eastman 910 Adhesive," literature by Eastman Kodak Company, 1958.
4. Loctite Sealants Corporation, East Hartford, Connecticut.
5. Twiss, S. B. "Adhesives of the Future." In *Symposium on Structural Adhesive Bonding*, edited by Michael J. Bodnar, held at Stevens Institute of Technology, Hoboken, N.J., September 1965. New York: Interscience Publishers, 1966.

CHAPTER 7

Adhesive Sealants

Sealants are not usually considered as adhesives and, conversely, adhesives are not usually discussed as sealants. Yet one of the advantages of adhesives, properly used, is that they very effectively seal a joint, as in aircraft fuel tanks and structural panels. Certain materials usually classified as sealants do a very good job of bonding structures, especially assemblies involving materials with widely differing coefficients of expansion. Therefore, this short chapter is included to describe three types of sealants which are being used more and more in bonding applications: polysulfides, polyurethanes, and silicones. Polyurethanes could have been included under adhesives (see Fig. 5-8). However, since they are elastomeric in nature they are also included in this chapter.

Because of their low moduli, elastomeric materials fall naturally into the role of stress relievers and thus fit in well with Chapter 8 on stress relief.

The materials discussed in this chapter are those which are available as castable, relatively solvent-free rubbers. It is in this form that they can be used as stress-relief interlayers and as vibration-damping mounts.

POLYSULFIDES

History

Polysulfide sealant development was initiated in 1932 with Thiokol Corporation acquiring sole rights to the patents.[1] Thiokol Corporation and the term polysulfides have been so closely associated ever since that the terms Thiokol and polysulfides are often used interchangeably.[2] The origins of polysulfide sealants are closely linked with commercial airplane development. As mentioned in the introduction, the DC-3 (the workhorse of the airlines), which was first built in 1936, used polysulfide sealants in the fuel tanks. As this chapter was being written, Trans Texas Airlines announced the retirement of their last DC-3 after 64,310 hours of commercial flying time and over 11,000,000 air miles under extreme conditions of exposure to temperature, vibration, fuels, and moisture.[3] This DC-3, designated as No. 797, was built in 1940 and served as a military transport in World War II before being used as a commercial airplane, starting in 1949. No. 797 will probably continue flying with uncounted other DC-3's all over the world, their fuel tanks still successfully sealed with polysulfides. This application is probably the outstanding example of the durability possible in a synthetic resin.

The polysulfide sealant formulations used in 1936 are no longer in use today, since materials and specification requirements have been improved. However, there is no reason why the present formulations should not be equally durable.

Formulation

The starting materials for the synthesis of polysulfide liquid polymers are bis-chloroethyl-formal and sodium disulfide. Liquid polymers with a molecular weight of 4000 form the basic ingredients of sealant formulations. These are cured primarily with metal peroxides. A representative reaction is:

Let

$$HS(-C_2H_4OCH_2OC_2H_4SS-)_{23}$$
$$-C_2H_4OCH_2OC_2H_4SH = -RSH$$
$$2(---RSH) + PbO_2 \xrightarrow{H_2O} (-RSSR) + PbO + H_2O.$$

As mentioned in the chapter on theory, most polysulfide formulations contain about 5 percent plasticizer. This is present in the catalyst. A representative formulation for the catalyst would be lead dioxide, 50 percent; stearic acid, 5 percent; dibutyl phthalate, 45 percent. The catalyst is usually added to the compounded liquid polymer in the ratio of 15 parts per 100 of liquid polymer content. The remainder is made up of vulcanizing ingredients, reinforcing fillers, and adhesion promoters.

Polysulfides give outstanding service as sealants because of adhesion, rubbery characteristics, durability, and fuel resistance.

Characteristics

Polysulfides are useful as sealants and adhesives over a temperature range from $-57C$ $(-70F)$ to $135C$ $(275F)$. While they do stiffen at low temperatures, they still retain some elongation at −54C (−65F) as shown in Fig. 7-1.

The specimens tested in Fig. 7-1 consist of two aluminum plugs 1⅛ inches in diameter × ¾-inch thick, bonded together on the 1⅛-inch diameter face with a 0.1-inch-thick layer of polysulfide.

The good adhesion characteristic of polysulfides is coupled with a toughness which does not propagate flaws. Polysulfides also tend to take a permanent set under load. These properties make them ideal for stress-relief applications.

Strength properties are affected by time and temperature, but not excessively, as shown in Table 7-1. The material always retains its tough rubberiness and adhesion and does not become brittle. The polysulfide evaluated in Table 7-1 was representative of sealants per MIL-S-8802, Class B2. The catalyst was manganese dioxide (MnO_2).

The data in Table 7-1 represents the results of three years' exposure to five different environments. The outdoor specimens were mounted at a 45° angle, facing south. The laboratory and semi-arid outdoor environment differ very little, with deflection being about the same at the end of three years.

The specimens exposed continuously in an oven at 74C (165F) changed the most in terms of deflection in shear, with 0.2 inch as a final value. The marine climate was most severe, causing considerable corrosion at the steel polysulfide interface for the three-year specimens. The paint did not protect the bond. For a similar series, in which a lead peroxide catalyst was used, paint did a much more effective job of protecting the bond. Continuous high humidity in the absence of sunlight rusted the steel but did not penetrate the interface. Except for the corroded steel-polysulfide interface all failures were cohesive in the polysulfide. The bond to aluminum never showed any sign of corrosion.

The picture (Fig. 7-2) of the bonded and soldered connectors demonstrates the ability of a resin bond to prevent electrolytic corrosion. Fifty hours of salt spray did not affect the polysulfide bonded connectors, while soldered joints were completely corroded.

Applications and Specifications

A measure of the effectiveness and adaptability of polysulfides is the number of specifications written on them as new applications were developed. Polysulfide formulations can be varied consider-

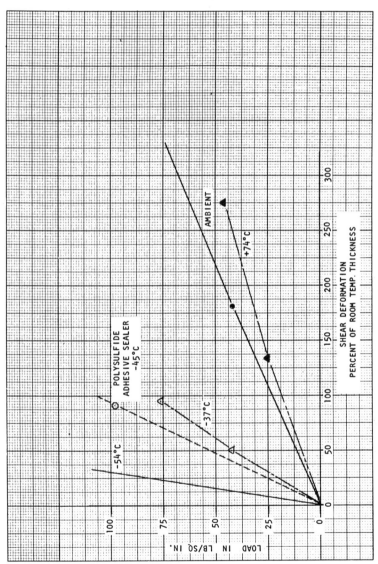

Fig. 7-1.   Load deflection for polysulfide adhesive sealer.

Table 7-1. Aging Characteristics of Sealant per MIL-S-8802, Class B2 (MnO$_2$ Cure) Shear Strength (S)[a], and Deflection (D)

Storage Site	Condition and Exposure	Initial (S) psi	(D) in.	6 Months (S) psi	(D) in.	12 Months (S) psi	(D) in.	3 Years (S) psi	(D) in.
Lab (ambient)		229	0.35	260	0.34	271	0.34	260	0.37
Albuquerque, New Mexico (semi-arid)	Painted[b]								
	Sun			196	0.44	198	0.41	226	0.36
	Shade			212	0.44	224	0.37	208	0.37
	Unpainted								
	Sun			211	0.44	206	0.41	254	0.35
	Shade			214	0.45	242	0.39	249	0.35
Kure Beach, North Carolina (marine climate)	Painted								
	Sun			185	0.40	162	0.35	110[c]	0.29
	Shade			181	0.42	188	0.37	104[c]	0.30
	Unpainted								
	Sun			185	0.43	170	0.36	100[c]	0.22
	Shade			184	0.43	185	0.35	104[c]	0.28
74C (165F) (oven)	Unpainted			307	0.26	308	0.24	279	0.20

Humidity[d]		10 Weeks (S) psi	(D) in.	6 Months (S) psi	(D) in.	10 Months (S) psi	(D) in.	12 Months (S) psi	(D) in.
	Painted	156	0.41	192	0.41	200	0.43	214	0.36
	Unpainted	158	0.42	210	0.42	206	0.41	212	0.37

a Each specimen consists of 1⅛-inch diameter plugs ¾-inch long, one aluminum, and one cold-rolled steel, bonded to each other on the 1⅛-inch diameter surface. The bond is loaded in shear, and deflection is measured in inches.
b Specimens were painted with air-drying olive-drab lacquer to simulate usual sealed joint.
c About 50 percent corroded at steel interface.
d Humidity cycle consists of not less than 93 percent RH with temperature cycling from 32C (89F) to 65C (149F) every 48 hours.

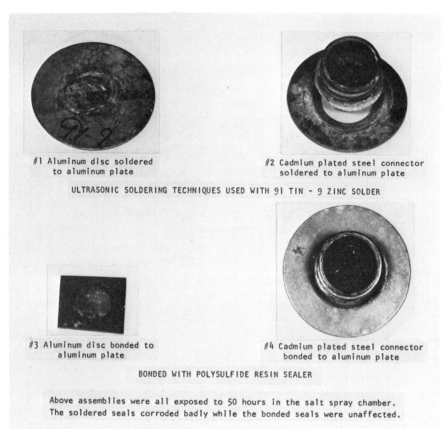

#1 Aluminum disc soldered
to aluminum plate

#2 Cadmium plated steel connector
soldered to aluminum plate

ULTRASONIC SOLDERING TECHNIQUES USED WITH 91 TIN - 9 ZINC SOLDER

#3 Aluminum disc bonded to
aluminum plate

#4 Cadmium plated steel connector
bonded to aluminum plate

BONDED WITH POLYSULFIDE RESIN SEALER

Above assemblies were all exposed to 50 hours in the salt spray chamber.
The soldered seals corroded badly while the bonded seals were unaffected.

Fig. 7-2.   Bonded and soldered connectors.

ably to meet specific applications. As the applications were formal-
ized specifications were written to cover them.

One of the oldest applications is as a sealant for integral fuel
tanks in aircraft. In this application the sealant must maintain
good adhesion and fuel resistance from $-54C$ ($-65F$) to $60C$
($140F$) in a vibration environment. The sealant must also contain
a minimum of extractable materials. It is interesting to note that,
because of the name, polysulfides are often unjustly suspected of
being corrosive because of their sulfur content. While free sulfur
is sometimes added in the amount of 0.1 percent or less in some
heat-resistant formulations, most formulations contain only the
chemically-linked sulfur present in the liquid polymer. The specification covering fuel-tank sealants is MIL-S-7502. This specification

calls out a 180-degree peel test using a polysulfide-coated fabric bonded to an aluminum strip, in addition to the usual shear specimens.

An upgraded version of MIL-S-7502 is the more recent MIL-S-8802. This has more rigorous requirements in that resistance to a temperature of 135C (275F) is included. Formulations meeting this specification include manganese dioxide and dichromate curing systems in addition to the usual lead-dioxide systems. A few of the specification requirements for polysulfides are given in Table 7-2. The specifications should be checked for details on environment and performance.

Table 7-2. Strengths and Volume Resistivity of Polysulfides

Polysulfide Specification	Shear	Elongation	Peel	Volume Resistivity, (ohm/cm)	
	Strength				
	psi	percent	pli	25C (77F)	65C (185F)
MIL-S-8516	. . .	. . .	15	2×10^{10}	2×10^{10}
MIL-S-8802B	200	200	20	. . .	. . .

The versatility of the polysulfides is illustrated by MIL-S-8784, which calls for a low adhesion sealant in applications where the strength of the bond is not critical. For these applications the sealant is required to strip easily with a two-pound pull, leaving no residue. This material is very useful for formed-in-place gaskets for access covers, and O-ring type seals.

Electrical properties for the usual polysulfide sealants are not exceptional—volume resistivities are about 10^{10} ohm/cm. With proper formulations, which emphasize nonpolar, reinforcing fillers and solvents, electrical properties are improved significantly with volume resistivity being about 10^{12} ohm/cm. The specifications for electrical grade polysulfide sealants are MIL-S-8516 and MIL-I-16923c.

These materials are used as connector sealants, as resilient conformal coatings on printed circuits, and as encapsulants.

The ability of the polysulfides to maintain seals and bonds in structural joints which are subject to wide ranges in expansion and contraction due to ambient and seasonal temperature changes has resulted in the writing of many specifications for bonding material such as methyl methacrylate and aluminum (MIL-S-7126A); for curtain-wall construction involving bonds to aluminum, glass, steel, and stone (ANSI Specification A116.1–1960); and for sealing formulations for exterior metal and wood construction (MIL-S-14231B, MIL-S-7124, and MIL-C-18255).

Structural bonds between materials of widely differing coefficients of expansion are sometimes possible only with adhesives such as polysulfides. Other means of fastening, such as: rigid epoxies, welds, or rivets, especially where large areas are involved, would build up sufficient stress to result in rupture either in the adhesive bond, in the adherend, or in both adhesive and adherend.

Processing

Polysulfides are usually available as 2-part kits which require weighing and mixing. These are sometimes put up in single units where the catalyst is pushed out into the resin. The unit, usually a polyethylene cartridge, is supplied with a push-pull mixer. Upon completion of the mixing, the cartridge is inserted into an air gun or hand gun for sealant application.

Polysulfides are provided as thixotropic pastes which can be applied to vertical or overhead surfaces, or as viscous self-leveling fluids which facilitate fill and drain operations or which can penetrate otherwise inaccessible voids.

One-part polysulfide sealants have been developed recently which simplify application on construction sites where weighing and mixing operations would be inconvenient. These depend on atmospheric moisture for curing and usually require several days to cure. One-part polysulfides are also described in Chapter 6.

POLYURETHANES

History

Polyurethane resins were developed in Germany about 1943.[4] Once introduced into this country after World War II, they gave great promise of being the "super rubber" which would result in indestructible products such as the 100,000-mile tire. Because of price, insufficient heat-resistance, and sensitivity to moisture, the potential has not been realized, but the promise is still there. Because of their excellent abrasion resistance, the polyurethanes are hard to ignore. They are widely used as flexible and rigid foams and are coming into their own as coatings. Isocyanate derivatives are also being used as primers for improving adhesion to woven fiber products.

Chemistry

The chemistry of polyurethanes includes amines and polyglycols reacted with isocyanates.[5, 6] Representative reactions are as follows:

(1) O = C = N⌒⌒⌒N = C = O + H – N⌒⌒⌒N – H $\xrightarrow{\text{FAST}}$

Diisocyanate   Diamine   Urea

(2) O = C = N⌒⌒⌒N = C = O + H – O⌒⌒⌒O – H $\xrightarrow{\text{SLOW}}$

Diisocyanate   Diol   Urethane

(3) O = C = N⌒⌒⌒N = C = O + H – O⌒⌒⌒O – H →

Diisocyanate   .   Triol   Urethane

While some of the reaction products are more properly known as ureas, it is accepted practice to call all materials of this type, polyurethanes.

Characteristics

Polyurethane adhesives have already been mentioned briefly in Table 3-9 to illustrate the great improvement in durability when used with a good primer. A polyurethane adhesive is included in Fig. 5-8 to show how well it performs as an adhesive at cryogenic temperatures. As shown in Table 3-9, when used alone without a primer, polyurethane adhesives can be sensitive to humidity. Bonding applications must take this sensitivity into account if they are to be successful.

Figure 4-15 uses a polyurethane adhesive to illustrate peel-strength dependence on bond thickness for a rubbery adhesive.

The potential strength and stress-relief capability of polyure-thanes as adhesive materials is shown by the material properties in Table 7-3 where two different prepolymers are combined, in vary-ing proportions, with a diamine cure. These data demonstrate the excellent strength, elongation, and tear strengths with no modifiers.

T-peel strengths of 300 pli for castable polyurethanes with the durability shown in Table 3-9 emphasize the toughness inherent in this material.

Table 7-3. Physical Properties of Polyurethane Rubber Formulations

Polyurethane Rubber Formulations[b]						
Materials	A	B	C	D	E	F
Adiprene L 100 (4.1% isocyanate)[a]	100	90	80	70	60	50
Adiprene L 213 (6% isocyanate)[a]	0	10	20	30	40	50
Methylene-bis-orthochloroaniline (MOCA)	11	12.8	14.6	16.5	18.2	20

Formulation	Test Temp	Tensile Strength, psi	Elongation, percent
A	−54C (−65F)	7,140 (5)*	240
B	−54C (−65F)	7,630 (5)	230
C	−54C (−65F)	8,000 (4)	240
D	−54C (−65F)	8,150 (4)	225
E	−54C (−65F)	8,650 (4)	80
F	−54C (−65F)	10,130 (4)	75
A	23C (73F)	3,390 (4)	450
B	23C (73F)	4,410 (4)	415
C	23C (73F)	4,380 (4)	370
D	23C (73F)	5,170 (4)	370
E	23C (73F)	5,210 (4)	300
F	23C (73F)	5,860 (4)	295
A	74C (165F)	1,550 (5)	300
B	74C (165F)	1,810 (3)	330
C	74C (165F)	2,250 (4)	360
D	74C (165F)	2,550 (4)	370
E	74C (165F)	2,680 (4)	320
F	74C (165F)	2,980 (4)	320
E	121C (250F)	1,140 (3)	235
E	177C (350F)	315 (3)	210

* Figures in parentheses indicate number of specimens tested.
[a] Adiprene is a trademark of the E.I. duPont de Nemours & Co., Inc.[5]
[b] Cure temperature was for 16 hours @ 74C (165F).

The use of polyols as chemically-linked modifiers improves processing by providing lower viscosities and longer working-life. While formulating with polyols tends to reduce tensile strength, the polyols in some formulations improve bond-durability when used with the right primers.[7] Table 7-4 gives properties for some amine/polyol cured formulations.

These data show that the use of polyurethane resins can provide a combination of strength and elongation not found in any other rubbery materials. Strength properties will vary depending on the type of polyol and amine/polyol ratio.

Applications

Polyurethane rubbers are available as solvent-free liquids curable at room temperatures or at intermediate temperatures up to 93C (200F). They can thus be cast in place, acting as both an adhesive and as a space filler. They are used to coat metal wheels and rollers and as vibration-damping mounts of unequalled toughness.[6] They can also be used where maximum strength is required in bonds between materials of different coefficients of expansion. However, since these bonds are sensitive to moisture, primers must be used.[8] At present these primers, available from different manufacturers,[9] must be cured at temperatures ranging from 93C (200F) to 121C (250F), and the polyurethanes must also be cured at elevated temperatures to obtain the best results. This limits the usefulness of the polyurethanes, but it is possible that work being done with silicone adhesion promoters may reduce the temperatures required to obtain optimum results.

SILICONES

History

Silicones are evidence of what concentrated development research can do with a new product. Silicones are available—both as one-part and two part fluid formulations and as thixotropic, nonflow sealants—for use as: water repellents; adhesion promoters; release agents; lubricating oils; greases; damping fluids; varnishes; solvent-type adhesives; molding compounds; vulcanized rubbers; defoamants; foam additives; cosmetics; car polishes; and room-temperature, vulcanizing (RTV), castable rubbers. The castable and thixotropic RTV rubbers are the materials with which this chapter is primarily concerned. Even in these few of the many silicone prod-

Table 7-4. Physical Properties of Some Polyurethane Formulations with Amine/Polyol Cures

Formulation[a]	Concentration	Ultimate Tensile[b], psi	Ultimate Elongation[b], percent	Secant Modulus to 10% Elongation[c], psi	Secant Modulus to 100% Elongation[c], psi	Shore "A" Hardness		Graves Tear-strength[d], pli
						Inst	10-sec delay	
Amines								
1. 1.00 eq Ad-L100 0.82 eq MOCA	100 gms 11 gms	3430	340	2600	950	87	85	305
Amines Plus Polyols Triols								
2. 1.00 eq Ad-L100 0.60 eq 11–80 0.33 eq MOCA	100 gms 14 gms 4.5 gms	1240	340	1270	445	77	74	145
3. 1.00 eq Ad-L100 0.40 eq 11–200 0.48 eq MOCA	100 gms 36 gms 6.6 gms	1490	460	1375	450	74	70	165
4. 1.00 eq Ad-L100 0.40 eq P-750 0.48 eq MOCA	100 gms 15 gms 6.6 gms	2750	565	1820	540	77	73	250
5. 1.00 eq Ad-L 0.30 eq P-1200 0.57 eq MOCA	100 gms 18 gms 7.7 gms	2690	600	1870	555	79	74	250

NOTE: [a] All specimens were cured for 16 hours at 104C (220F).
[b] ASTM D412-51T.
[c] ASTM D638-52T.
[d] ASTM D624-54, Die C.
AD-L100 = Adiprene L100 produced by E.I. duPont de Nemours & Co., Inc.
11–80, 11–200, P-750, and P-1200 = Polyglycols produced by The Dow Chemical Co.
MOCA = Methylene-bis-orthochloroaniline produced by E.I. duPont de Nemours & Co., Inc.

ucts, such great strides have been made in the past few years in strength and processing characteristics that one would not recognize the present RTV rubbers as being related to the initial, RTV, castable, silicone rubbers.

The long history of silicon chemistry commenced bearing practical, commercial fruit about 1942, when both General Electric and Dow Corning (a company created by Dow Chemical and Corning Glass) decided to put concentrated effort into the development of heat-resistant resins based on silicones.[10, 11] By 1945, silicone rubber was a commercial fact.

Chemistry

The first RTV silicone rubbers were reaction products of hydroxy terminated, dimethyl silicones with organosilicates:

This type requires a certain minimum amount of water to initiate the hydrolysis of the silicate and start the reaction. It does not cure well in deep sections and may depolymerize in the presence of residual reaction products if exposed to high temperatures too soon after an initial room-temperature cure.

The vinyl terminated silicones represent an improvement on this system. They react to form an addition product with a hydrogen-methyl-silicone curing agent as follows:

This type cures well in deep sections, is independent of moisture availability, and can be exposed immediately to elevated temperatures. The cure is inhibited by sulfur-containing compounds and by metallo-organic catalyzed RTV systems. Another RTV-silicone innovation is the one-part room-temperature cured, silicone-rubber adhesive system. This type of adhesive contains chemically-blocked curing agents which are hydrolyzed by moisture contained in the air. One product of this hydrolysis is removed by evaporation while the other product proceeds with the cure.

The two-part silicones require adhesion promoters, usually dual-function silane primers, in order to bond to nonsilicone surfaces. The one-part silicones develop so-called "natural" adhesion on hydrolysis of the chemically-blocked curing agents, so that they may bond well without primers, but quite often give improved adhesion with primers. Volatile by-products of the one-part silicones may vary from possibly corrosive acetic acid to noncorrosive alcohol.

The one-part silicones are usually used as edge sealants because of the need for atmospheric moisture to achieve a cure. However, bonds between two impermeable adherends are made possible simply by spreading the one-part adhesive thinly over both the surfaces to be bonded, exposing these surfaces to the atmosphere while the process of moisture-absorption and curing-agent hydrolysis takes place. The time of exposure should be less than the "skin-over" time given in the manufacturer's literature. The surfaces to be bonded are assembled under sufficient pressure to achieve complete contact. The silicone resin then cures to a tough, resilient, stress-relieving interlayer.

Properties

Some of the properties of the high strength one-part and two-part, RTV silicones are shown in Table 7-5. The properties mainly are taken from the literature and represent the range of properties possible from the available materials.[12]

The one-part silicones differ from the two-part silicones primarily in cure conditions in that they are sensitive to relative humidity (curing faster at higher humidities), and cure rates are inversely proportional to thickness so that cures are not advisable for thicknesses greater than 1/4-inch. The one-part silicones also have a higher percentage of volatile material, as might be expected from the cure mechanism. However, once the bond is cured, vacuum baking is possible at temperatures ranging up to 204C (400F), so that volatile contaminants can be eliminated. Recent improvements have resulted in silicone RTV materials with a volatile content of less than 0.1 percent.[13] This sort of development has been spurred on by requirements of minimum volatility for outer space applications.

Table 7-5. Silicone-Rubber Properties

Properties	High-strength RTV Silicones,	
	One-Part	Two-Part
Spec gravity (cured)	1.07 to 1.12	1.07 to 1.20
Viscosity (poises)	Usually thixotropic	600 to 2000
Useful temp range, C (F)	−65 to 300C	−65 to 275C
	(−84 to 572F)	(−84 to 527F)
[a]Thermal cond 25 to 100C, (77F to 212F)	...	3.5–7.5×10^{-4}
Wt loss, 24 hrs at 200C (392F)	Up to 6 percent	0.1 to 1.0 percent
Vol. expansion (cc/cc/C)	...	8×10^{-4}
Specific heat (cal/gm/C)	...	0.34
Tensile strength (psi)	350 to 700	600 to 800
Elongation (percent)	300 to 600	300 to 400
Hardness (Shore "A")	25 to 35	30 to 60
Tear strength (pli)	45 to 100	80 to 100
T-peel strength (pli)	...	50 to 100
Deep section cure	No	Yes
Dielectric constant at 10^6 cps		
21C (70F)	...	3.13
94C (200F)	...	3.05
Dissipation factor at 10^6 cps		
21C (70F)	...	0.004
94C (200F)	...	0.010
Vol. resistivity (ohm/cm) 21C (70F)	...	10×10^{15}
94C (200F)	...	10×10^{15}
Dielectric strength (volts/mil)	...	500

[a] cal/cm²/sec/°C/cm

Fluorinated high-strength silicones, both one-part and two-part, are also available where solvent and fuel resistance are required, as in fuel-tank sealants for supersonic jets.[14]

The properties that make the more recently developed silicones different from earlier RTV silicones are tensile strength and elongation, coupled with high tear-strength and T-peel strength (see also Table 3-8). Earlier materials were notch sensitive so that flaws propagated rapidly under stress, and full strength was seldom realized. Recently developed RTV materials retain their strength under stress so that when used with the proper primers, full benefit is derived from their tensile strength and elongation. In addition, these materials are useful as rubbers at a wide temperature range from lower than −74C (−101F) to higher than 250C (482F). For short periods of time they are useful at high temperatures of over 400C (752F).

Stress Relief Capability

The polysulfides and polyurethanes already described had useful rubbery properties from −60C (−76F) to about 100C (212F).

However, the polysulfide measured in Fig. 7-1 differs by a factor of 20 in deformation for a given load over a temperature range of 100C (212F). The polyurethanes evaluated in Table 7-3 vary in tensile strength by a factor of almost 5 over the same temperature range. Essentially this means that over a temperature range of approximately 100C (212F), the stress-relief capability and vibration-damping characteristics vary considerably. For most applications this variation may not be important. However, in some applications uniformity of stress relief and vibration damping over a wide temperature-range is a necessity, and here silicone rubber is unequalled. Figure 7-3 gives an indication of this in terms of shear deformation under load over a 100C (212F) temperature range. While the silicone shows some change in this temperature range, it is very small compared to the changes in the polysulfide (Fig. 7-1) and the polyurethane evaluated in Table 7-3. Silicone RTV s prepared from a methyl-phenyl silicone base have useful rubbery properties at temperatures as low as −100C (−148F). Increase in the phenyl substituent also tends to improve the ablative properties of RTV silicones.

Structural Capability

Until a few years ago, RTV silicone rubbers were used primarily as encapsulants, as conformal coatings, or as sealants which, in addition to performing the sealing function, needed only to support their own weight. Present materials are considerably stronger than polysulfides, with peel strengths surpassing those of most of the structural adhesives if sufficient thickness is allowed in the bond (see Table 3-8 for reference). In addition, the RTV silicones are probably usable over a wider temperature-range than most, if not all, of the structural adhesives considered in the previous chapter; they also exhibit a greater usable temperature-range than any purely organic rubber.

Thus if the strength limitations are not exceeded, silicones can now be considered for structural members where stress relief and vibration damping are required. While other molded-rubber products are available for such applications, none of the carbon-base materials is useful over a similar temperature range. The advent of the supersonic era with its higher temperature requirements may result in considerable use of these RTVs.

The usual silicones are susceptible to swelling and degradation when wet with organic solvents, including petroleum fuels. However,

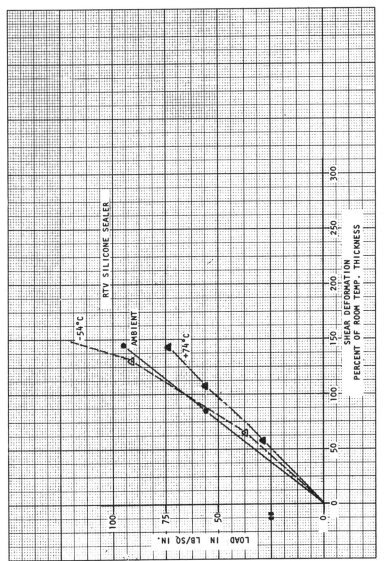

Fig. 7-3.　Load deflection RTV silicone sealer.

the addition of fluorine to the silicone polymer chain has improved fuel and solvent resistance, so that these materials are useful as sealants and adhesives in the presence of organic solvents at temperatures in excess of 150C (302F). The fluorinated silicones are the leading, if not the only, candidates being considered for use as fuel-tank sealants in the supersonic transports.

In contrast to the conventional, organic-rubber formulations which require expensive molds and hot-press equipment for processing, RTV formulations can be processed with relatively inexpensive casting molds. Castable polyurethanes and polysulfides would use similar, inexpensive tooling.

Manufacturers of silicone RTV systems are, primarily, Dow Corning Corporation of Midland, Michigan, and General Electric Company, Silicones Division, Waterford, New York. Other suppliers are Union Carbide Corporation and Linde Stauffer Chemical Company.

Prices for silicones are about $5 per pound, depending on quantity, and fluorinated silicones are much more expensive (about $50 per pound).

REFERENCES

1. Patrick, J. U.S. Patent 1,890,191 (1932).
2. Panek, J. R. "Polysulfide Sealants and Adhesives." In *Handbook of Adhesives*, edited by I. Skeist. New York: Reinhold Publishing Corp., 1962.
3. *Albuquerque Journal*, February 1, 1968.
4. Schollenberger, C. S. "Isocyanate-Based Adhesives," Ibid. 2.
5. Quant, A. J. *A Castable Polyurethane*, Sandia Corporation Reprint SCR-54, October 1958. Available from Office of Technical Services, Department of Commerce, Washington, D. C.
6. Saunders, J. H., and Frach, K. C. *Polyurethanes Chemistry and Technology*. New York: Interscience Publishers, 1964.
7. Athey, R. J.; DiPinto, J. G.; and Keegan, J. M. "Adiprene L100, A Liquid Urethane Elastomer." *DuPont Bulletin No. 7*, October 1965.
8. DeLollis, N. J.; Montoya, O.; and Curlee, R. M. *Effect of Assembly Variables on Peel Strength of Adiprene Bonds*. Sandia Corporation Report SC-DR-65-161, June 1965.
9. Dayton Chemical Products Co., Dayton, Ohio; Hughson Chemical Co., Akron, Ohio; Conap Inc., Allegheny, N. Y.
10. McGregor, R. R. *Silicones and Their Uses*. New York: McGraw-Hill Book Co., Inc., 1954.
11. Fordham, S. *Silicones*. New York: Philosophical Library Inc., 1961.
12. General Electric Corporation, Silicones Department, Waterford, New York; Dow Corning Corporation, Midland, Michigan.
13. "Process Gives Low Volatility Materials Aimed at Use in Space," *Chemical and Engineering News*, April 8, 1968, p. 24.
14. Harisis, H. G. "Development and Evaluation of a High Temperature Groove Injection Sealant." *SAMPE Journal*, April/May 1968, pp. 52-62.

CHAPTER 8

Stress Relief, Vibration Isolation, and Vibration Damping

Low modulus, elastomeric materials perform a unique service in modern civilization when one considers first, the multitude of materials of varying coefficients of expansion which must eventually be fastened together; and second, the vibration and other cyclic loadings which modern machines generate to add fatigue to the degradative environment which already includes such factors as heat, oxidation, humidity, and radiant energy.

The part that fatigue played in the history of the Comet, the first commercial jet aircraft, has already been described in the chapter on test methods. Recently, Boeing announced an extensive wing-modification program aimed at preventing potential fatigue problems in its 707 aircraft.[1] The fatigue problems in this instance are due to the cyclic loads imposed by landings, takeoffs, and taxiing. The results of these cyclic loadings are made evident as cracks around the wing fasteners, noticeable after about 25,000 flying hours.

The problem of fatigue is being given considerable attention, and adhesive bonding with and without mechanical fasteners is one of the means of minimizing this potential cause of equipment failure.

Adhesives such as nylon epoxies, nitrile-rubber phenolics, epoxies, and elastomers have low-modulus characteristics, and, when properly applied, allow machines to operate successfully in a vibration environment. Otherwise, machines such as automobiles and airplanes would soon shake themselves to pieces if exposed directly to vibration. Nowhere is this more strikingly illustrated than in the development of the helicopter rotor-blade. The modern rotor-blade is a completely bonded unit. The use of adhesives in place of rivets has increased the life-expectancy of the rotor-blade from 200 to more than 2,000 hours, by simply spreading the fatigue stresses over all of the contacting surfaces.[2]

Stress Relief

Another area where elastomeric adhesives efficiently minimize stress is the assembly of materials with greatly differing coefficients of expansion. Figure 8-1 charts the extremes to be encountered in materials for which modern technology may require an assembly method.

Unfilled organic synthetics are represented by nylon phenolic with a coefficient of expansion of 50×10^{-6} inch/inch/$^\circ$F. The nylon phenolics are very efficient ablators with good thermal insulation properties. This type of material is used to protect missiles and manned, orbiting vehicles from temperatures encountered on re-entry into the earth's atmosphere. When it is bonded to a structural metal case, obviously large stresses must result because of the great differences in coefficients of expansion even if the structural case is made of aluminum. The resulting stresses are proportional to the net difference in coefficients of expansion, to the cycling temperature range from the cure temperature, and to the moduli of the adherends.

In actual practice a 3-foot-diameter aluminum case, bonded to a nylon-phenolic, ablative shield with a room-temperature curing epoxy cracked the shield on cooling to -54C (-65F) and tended to fracture the bond in tension, on heating to 74C (165F). The bond line-thickness was about 0.005 inch. With the same design, but with a polysulfide sealant per MIL-S-7502, Class A, and with a bond line-thickness of 0.100, the assembly was cycled repeatedly from -54C (-65F) to 74C (165F) with no bond-failure. Voids were built into the adhesive layer to provide space to facilitate extension and compression of the polysulfide.

In another experiment, a carbonized, fiber-laminate ring was bonded to an aluminum ring with the aluminum ring on the inside representing a structural case. The diameter at the bond was 4.0 inches. Since a heat-resistant bond was needed, an epoxy-phenolic-adhesive curing at 300F was used. The adhesive bond ruptured on cooling to room temperature. For this combination a high tear-strength, RTV-silicone adhesive would have been much more successful. Scaling up to a full-size assembly would have required increasing the bond thickness to provide the necessary elongation to make up for the difference in coefficients of expansion. Such an assembly with an RTV-silicone adhesive, cured at room temperature, would allow temperature cycling with very little stress in the bond.

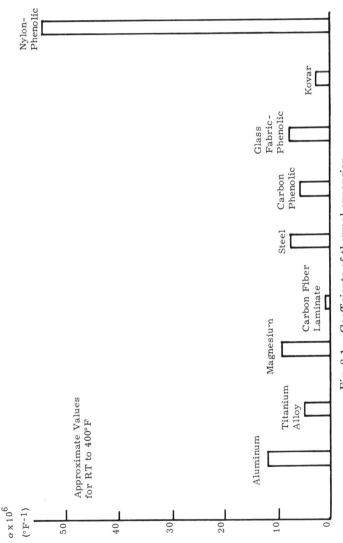

Fig. 8-1.   Coefficients of thermal expansion.

Occasionally, in bonding materials of differing coefficients of expansion, attempts are made to relieve stresses by using semirigid epoxy resins as the adhesive. These materials, however, tend to weaken excessively at elevated temperatures.

Experimental Results

In order to obtain a better understanding of the variables involved in bonding a nylon-phenolic heat-shield to an aluminum case, a series of tests was run using a small-scale model as shown in Fig. 8-2. The heat shield was nylon phenolic with a coefficient of

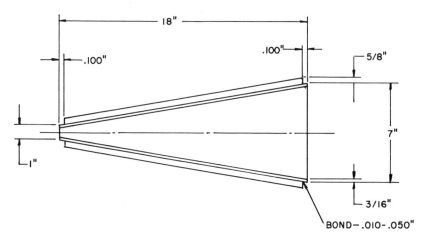

Fig. 8-2.   Experimental model.

expansion of 50–60 × 10⁻⁶ inch/inch/°F. Deflection readings were taken between the ends of the aluminum case and the ends of the nylon phenolic at the following temperatures and in the following sequence: Ambient, −37C (−35F), 54C (130F), −54C (−65F), 74C (165F), 93C (200F), 121C (250F), and 149C (300F). The assembly was conditioned in the temperature chamber for 3 to 4 hours and measured outside of the temperature chamber as soon as possible to minimize the effect of warming or cooling. The adhesives evaluated are described in Table 8-1.

The epoxy adhesives were cured 4 to 7 days at room temperature before the start of the test. The silicone-rubber foam was foamed and cured at 94C (200F). The bond thickness for the epoxy adhesives was about 0.010 inch, while for the foam it was about 0.050

Table 8-1. Type and Source of Adhesives Used in Figure 8-3

Adhesive	Type	Supplier
A Epoxy resin with 20 to 30 percent Bentone clay thickener (60 parts by weight) and polyamide curing-agent (40 parts by weight)	Semirigid	. . .
B Proprietary polysulfide-modified epoxy	Semirigid	B. F. Goodrich Co.
C Rigid epoxy with glass fiber and aluminum filler (proprietary)	Per MIL-A-8263, Type I	Shell Chemical Co.
D General Electric silicone-rubber foam adhesive No. 757	. . .	General Electric Co., Silicones Dept.

inch. The nylon-phenolic surface was sandblasted, then wiped with methyl-ethyl ketone. The aluminum was etched with a sulfuric acid, dichromate solution per MIL-A-9067. The results are shown in Fig. 8-3. The measurements at ambient were used as the reference point and were plotted as zero deflection. For the semirigid epoxies and the rigid epoxy, the deflections with temperature were very small. This indicated that, until the bond failed, the stresses were almost completely contained with very little strain. At the temperature where the stresses exceeded the bond strength, bond failure was evidenced at the narrow end with the edge of the heat shield advancing by as much as 0.2 inch at the highest exposure temperature—149C (300F) for the rigid epoxy.

The polyamide-cured epoxy bond failed when the temperature was raised to 74C (165F). The polysulfide-modified epoxy bond failed when the temperature was raised to 93C (200F). The rigid, epoxy bond successfully contained the stresses through the 93C (200F) and 121C (250F) exposure but failed when the temperature was raised to 149C (300F). The arrows in Fig. 8-3 indicate the temperature range in which the bond failed.

The silicone-rubber-foam adhesive essentially allowed the shield and case to expand independently of each other in a relatively stress-free condition, resulting in two smooth curves with no sudden break because of bond failure.

If one could ignore the change in modulus with temperature of the materials involved in this study, it is conceivable that the tem-

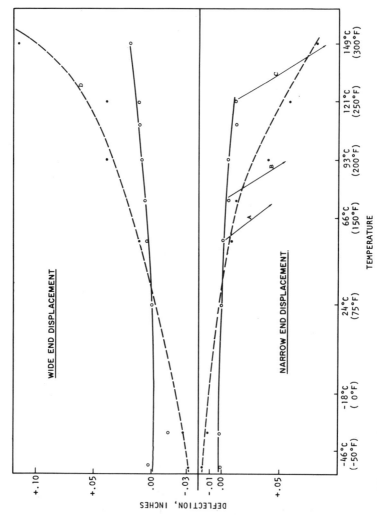

Fig. 8-3.   Longitudinal displacement of nylon phenolic heat shield from aluminum case

perature at which bond failure would occur in larger structures could be estimated by assuming that stress would increase linearly with size, and that failure would occur at a proportionally lower temperature. Thus a compressible, expandable, relatively stress-free elastomeric bond would be considered if the failing temperature for a rigid adhesive was too low.

The foregoing demonstrates the fact that stresses in the bond due to adherends of different coefficients of expansion can be contained by the adhesive to a certain extent, and they can be contained best by a rigid adhesive. At some temperature, these stresses will become too great to be contained so that the assembly will rupture either in the bond area or in one of the adherends. Once this point is determined, either by experiment or by structural analysis, the decision to use an elastomeric adhesive becomes unavoidable.

VIBRATION AND STATIC FATIGUE

The increase in useful life of a helicopter rotor-blade when the fastening method was changed from rivets to adhesives is an outstanding example of benefits to be derived from the use of adhesives in bonded structures.

Stresses, leading to fatigue failure in metal skins, can be due either to static loads, or, as in the case of a pressurized aircraft, to continuous vibration imposed on a static load. Probably this sort of fatigue was mainly responsible for failure of the Comet aircraft described in the chapter on test methods.

DeBruyne, in describing some of the first work in which the fatigue life of a bonded joint is compared to that of a riveted joint, stated that in a properly-bonded joint the fatigue life approached that of the metal itself.[3] Some of the work reported at that time indicated that the bonded joint, with regard to fatigue, was better than the metal itself and more than six times better than the riveted joint.

In double lap-shear joints (reported in Reference 4) prepared from 0.064-inch aluminum strips bonded to 0.032-inch strips with a vinyl-phenolic (Redux) adhesive, both static and fatigue tests resulted in failure in the metal rather than in the bond with a metal stress of 32.8 tons/inch.[2] The static results are similar to those obtained with the double lap-shear tests as described in the chapter on joint design, Table 4-2.

As shown in Fig. 8-4, the bonded, double lap-shear joints with-

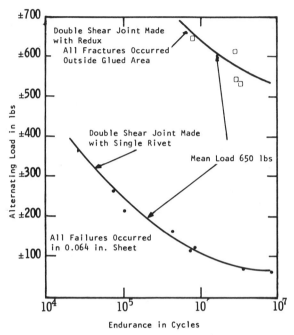

Fig. 8-4.   Comparison of fatigue life of double lap joints made by "Redux" bonding and by riveting.

stand six times the shear load of the double lap-shear joint made with a single rivet. All failures in the bonded joint occurred outside the bond area.

Crack Stoppers

Cracks in metal due to fatigue are more effectively arrested by bonded crack-stoppers than by riveted crack-stoppers. Figure 8-5, taken from an NACA report, shows a comparison of crack-propagation rates in aluminum alloy box beams with riveted, bonded, and integrally machined stiffeners.[5] Bonded reinforcements in riveted joints and bonded stiffeners have also been used with success in primary structures subject to fatigue.

In earlier metal aircraft construction, bonded seams were terminated with rivets to minimize the possibility of stresses initiating peeling at the bond ends. However, with the development of durable tough adhesives such as nitrile phenolics and nylon epoxies, this is no longer necessary. The B-58 bomber is an ideal example of bonded aircraft construction.[6]

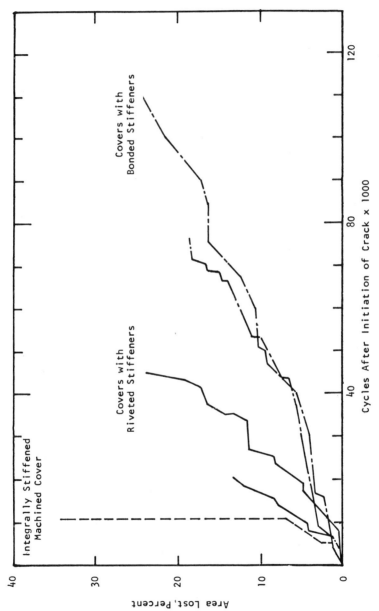

Fig. 8-5. Crack propagation in machined, riveted, and bonded panels.

Bonds with Rivets

Because of the size of wing or body panels, it is sometimes necessary to use a combination of rivets and adhesive bonding. The rivets in this sort of application double as fasteners and as holding and pressure-tooling for the adhesive bond. Table 8-2 taken from Reference 7 shows the improvement to be expected from this combination.

Table 8-2. Fatigue-testing Results of Longitudinal,
Lap-splice Test Specimens*

Configuration	Cycles to Failure[a]	Failure
No bond	211,000	In first row of rivets in countersunk sheet
Epon bonded[b]	1,500,000	No failure
Epon bonded[b]	1,500,000	No failure
Pressure-sensitive tape	33,000	In first row of rivets in countersunk sheet
Thiokol bonded	42,000	In first row of rivets in countersunk sheet
Machined pad on skins—no bond	230,000	In first row of rivets in countersunk sheet
Chem-milled pad on skins	180,000	In first row of rivets in countersunk sheet
Doubler bonded to both skins with nitrile phenolic	212,000	In first row of rivets in countersunk sheet

* Taken from Reference 7.
[a] Stress level 8000 ± 7700 psi for room-temperature tests.
[b] Tested to 1,500,000 cycles. Rivets removed and riveted, stressed to 19,500 psi at 82C (180F), reduced to −54C (−65F) under stress, and recycled.

It is interesting to note that the nitrile-phenolic adhesive-rivet combination did no better than rivets alone. The explanation probably lies in the fact that when combined with rivets, the adhesive must exhibit minimum strain so that no localized stress results at the rivet-skin contact. A low-modulus adhesive such as the nitrile phenolic, which works very well by itself, has sufficient elongation in the cured state so that the applied load is transferred to the rivets

The pressure-sensitive tape and the polysulfide (Thiokol)-rivet combination were considerably worse than rivets alone. The explanation here is probably that they contributed to a certain bond-thickness at the rivet joint, thereby providing considerable room for play in the joint, while contributing no support at all because of their very low modulus.

The British report that a bonded, riveted joint has a longer fatigue-life than a riveted joint, when room-temperature-curing adhesives are used.[8]

VIBRATION-DAMPING MOUNTS

Wherever a source of power causes excessive vibration, damping devices are used; for example, vulcanized-rubber mounts in automobile engines. The development of castable room-temperature-curing elastomers such as the polyurethanes, polysulfides, and silicones makes it possible to design vibration-damping and vibration-isolating mounts with a minimum of equipment and with a wide choice of elastomer material properties. However, any vibration-damping designer requires detailed information as to the modulus variation of the elastomers with temperature and frequency. This information is quite often not available.

Temperature Dependence

Figure 8-6 (from Reference 9) shows how modulus varies with temperature and frequency for a polysulfide sealant per MIL-S-7502, Class B. This material in a metal laminate has good vibration-damping properties, but the modulus is sensitive to temperature and frequency. Anyone planning to make use of it must be sure that the temperature and frequency ranges of the application do not exceed the capability of this polysulfide resin.

Table 8-3 shows a comparison of the moduli of the polysulfide sealant with those of castable polyurethane rubber and a high-strength, castable silicone rubber for three different temperatures. Of the three materials, the polysulfide is the weakest at ambient and at 74C (165F), and is the most sensitive to temperature, exhibiting a change in modulus by a factor of about 20, from the highest to the lowest temperature. The polyurethane is the strongest and is somewhat less sensitive to temperature, with a modulus change by a factor of 6, from the highest to the lowest temperature. The modulus values for the silicone elastomer emphasize the uni-

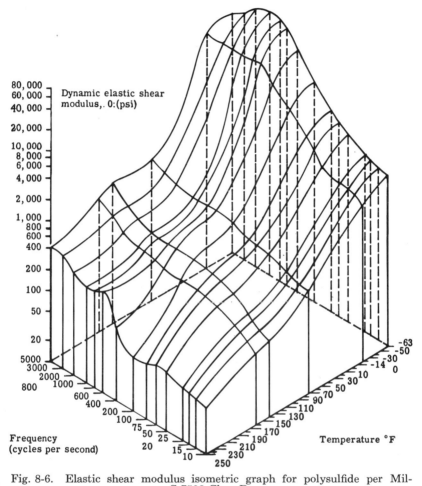

Fig. 8-6.   Elastic shear modulus isometric graph for polysulfide per Mil-S-7502 Class B.

Table 8-3. Modulus versus Temperature for Four Synthetic Elastomers

Temperature	Nitrile Phenolic[a]	Polyurethane Castable Rubber[b]	Polysulfide per MIL-S-7502[a]	Silicone RTV[b]
	pounds per sq in.			
—54C (—65F)	. . .	2970	320	159
24C (75F)	2000	750	25	136
74C (165F)	. . .	520	17	148

[a] Shear modulus.
[b] Tensile modulus at ultimate elongation.

formity of silicone elastomer properties over a temperature range, with essentially no change in modulus, from $-54C$ ($-65F$) to $74C$ ($165F$).

While detailed information is not available for the polyurethane and for the silicone, it is obvious that the polyurethane would be used where greater strength was required, while a silicone elastomer would be used where uniformity over a temperature range was the governing requirement.

One modulus value is included from Table 4-1 to show how these moduli compare with the modulus of a nitrile-phenolic resin, the most rubbery of the conventional structural adhesives.

Composite Construction

Adhesives such as the nitrile phenolic improve the fatigue life of a metal structure in a vibration environment by spreading the load stresses over all of the contacting surfaces; in contrast, stress in a riveted joint is highly localized. An added function of the adhesive is to form a sandwich with the two metal layers at the bonded joint. The sandwich has a lower average modulus than the straight metal skin, which results in a damping action at the resonant frequency or frequencies.

Robertson describes one practical application of a sandwich construction in which two ballistic cases were built in order to compare the damping characteristics of a metal-polysulfide laminate with those of a solid-metal wall-construction.[9] The outer metal shell of the sandwich construction was attached to a mass forming the nose, while the inner metal shell was attached to the vibration support fixture. When the case was vibrated in the manner of a cantilever beam, the polysulfide was strained by shear forces, and it dissipated energy as it was strained cyclically. Both cases were vibrated over a frequency range of 0 to 2,000 cps. The major resonant frequency occurred at about 120 cps; the maximum amplification for the undamped case was 140, while the maximum amplification for the damped case was 25.

In the usual course of events the equipment package to be exposed to a vibration environment is designed independently of its vibration-protection needs. It is then necessary to design a mount which will isolate the package from the vibration environment. If the input frequency from the vibration source coincides with the natural frequency of the equipment package, then a resonant condition develops. In the experiment with the ballistic case described

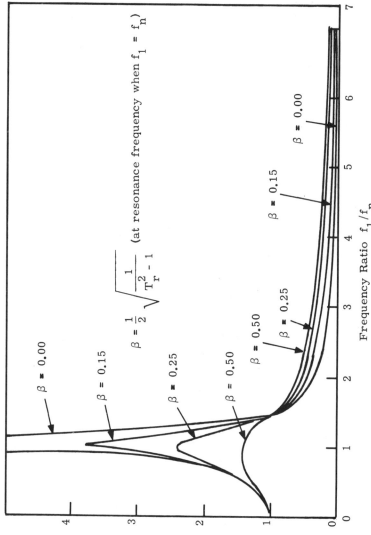

Fig. 8-7. Vibration response of a damped single-degree-of-freedom spring-mass system with various damping factors.

above, such resonance resulted in an amplification factor of 140. Vibration isolation requires that the natural frequency of the isolation system be well below the input frequency.[10, 11] This results in a damping action at the resonant frequency, depending on the damping factor of the system. However, at frequencies other than the resonant frequency an increase in transmissibility occurs, as is shown in Fig. 8-7.[10] This increase in transmissibility must be taken into account to insure that it is acceptable.

Vibration Isolation

Another approach to the problem of objectionable resonant-frequency would be to redesign the package to change its resonant frequency so that it did not match the input frequency. This could be accomplished by replacing structural members such as solid-sheet stock with laminated-sheet material with a lower modulus than the solid sheet. These are available commercially.[12] At the same time, vibration-damping mounts could be designed into the unit.

Table 4-1 and Table 8-3 demonstrate the wide range of moduli available in structural adhesives and in adhesive sealants. Properly used, they can contribute significantly to increasing the life of metallic structures exposed to a cyclic fatigue or vibration environment. The adhesives shown in Table 8-1, when improperly used, can be harmful. Adhesive sealants with good elongation, good adhesion, and high tear-strength allow the assembly of different materials with no limitations based on difference in coefficients of expansion.

These adhesives and adhesive sealants are available as solutions, pastes, films, and solventless fluids. Cures can be expected at temperatures ranging from ambient, to intermediate, to high (175C [347F]). All of these options in processing and cure mean that these materials can be adapted to fit in with practically any production process. The principal requirement is that materials engineers be fully acquainted at all times with the latest developments in synthetic resins including adhesives and adhesive sealants.

REFERENCES

1. O'Lone, R. G. "Airlines Begin 707 Wing Program." *Aviation Week and Space Technology*, February 5, 1968, pp. 32, 33.
2. Bair, F. H. "Space Age Adhesives." *Adhesives Age.* 2 (1959): 20–22.
3. DeBruyne, N. A. *Redux in Aircraft.* New York: Ciba Co. Inc., 1953.

4. DeBruyne, N. A. "Joint Design for Primary Structures." In *Symposium on Adhesives for Structural Applications.* New York: Interscience Publishers, 1962.

5. Hardrath, H. F. et al. "Fatigue Crack Propagation in Aluminum Alloy Box Beams," National Advisory Council in Aeronautics Technical Note 3856, 1956.

6. Lunsford, L. R. "Design of Bonded Joints." In *Symposium on Adhesives for Structural Applications.* New York: Interscience Publishers, 1962.

7. Hilton, J. R. "Use of Room Temperature-Curing Adhesive Film on the 727 Airplane." In *Applied Polymer Symposia on Structural Adhesive Bonding,* edited by Michael J. Bodnar. New York: Interscience Publishers, 1966.

8. Catchpole, E. J. "Some Recent European Developments in the Structural-Adhesives Field." In *Applied Polymer Symposia on Structural Adhesive Bonding,* edited by Michael J. Bodnar. New York: Interscience Publishers, 1966.

9. Robertson, A. B. *Properties and Applications of Viscoelastic Materials.* Sandia Corporation Technical Memorandum SC-TM-65-535, December 1965.

10. Rand, P. B. *Vibration Isolation of Systems Under Quasi Steady State Accelerations with Flexible Polyurethane Foam.* Sandia Corporation Technical Memorandum SC-TM-66-340, July 1966.

11. Thomson, W. T. *Vibration Theory and Applications.* Englewood Cliffs, N. J.: Prentice-Hall, Inc., 1965.

12. Barry Controls, Inc., Watertown, Massachusetts; Lord Manufacturing Co., Erie, Pennsylvania; and Minnesota Mining and Manufacturing Co., St. Paul, Minnesota.

CHAPTER 9

Industrial Processing of Adhesives and Bonded Assemblies

FASTENING METHODS

Industrial processing of metals in preparation for assembly to other metals assumed a new dimension when adhesives entered the picture. Rivets and nuts and bolts used as metal fasteners are concerned primarily with the bulk properties of the metals. Surface treatments such as pickling, priming, and painting are primarily corrosion preventives. In some instances a thin film of oil or grease suffices in the battle against corrosion. These coatings neither hurt nor help the strength of the mechanically-fastened joint.

Soldering and brazing may be said to relate to adhesive bonding since they rely on surface attachment. Surface preparation is important since the presence or absence of oxides, grease, and dirt determine the extent to which solder or brazing-metal wet and bond to the base metal. However, it is also required that the base metal be heated above the melting point of the solder or brazing-metal so that the latter, with the help of suitable fluxing agents, may wet the metal and run freely. At this point the use of molten metal as an assembly device more closely resembles the coalescence of thermoplastic materials being fused together by heat than it does the structural adhesive-bonding of metals, since a certain amount of alloying does take place.[1] The cooled and fully-hardened joint usually has a transition area of diffused metal, rather than a distinctive interface.

Welding requires less surface processing than either soldering or brazing because welding temperatures are those which melt the metal being joined. Here the bulk behavior of the molten metal is important.

In contrast to dependence on bulk behavior and on metal melting-temperature, adhesive-bonding technology places as much emphasis on surface preparation as on adhesive application and cure.

Surface preparation has already been covered in Chapter 3. This chapter is concerned with adhesive application, assembly tooling, and processing equipment, in order of complexity. Just as cleaning practice can vary from simply wiping with a solvent-soaked cloth to elaborate etching, so adhesive practice may vary from curing one-part adhesives at room temperature with practically no holding fixtures, to curing adhesives under pressure in giant autoclaves with the use of elaborate holding fixtures.

SIMPLEST PROCEDURE

One-part, Room-temperature-cure Adhesives

The simplest bonding practice would be followed with one-part adhesives which would cure at room temperature with little or no bonding pressure. As a concession to positive positioning to minimize joint-slippage during cure, simple spring clamps, C-clamps, or possibly dead-weight loading should be permissible.

Use of one-part adhesives has obvious advantages. It eliminates the risk entailed in weighing two or more ingredients and in mixing these ingredients uniformly. Weighing requires accurate scales which should be calibrated periodically to insure accuracy. Mixing might require equipment which would need cleaning. Many a production crisis has involved the use of the wrong curing agent or the weighing of improper amounts of adhesive components. To minimize such costly mistakes, detailed process specifications are written and day-to-day laboratory notebooks are kept, to note materials weighed and weights made. The use of room-temperature cures eliminates the need for ovens, heat lamps, heat pads, and other sources of heat which, while not necessarily expensive, do require some expenditure and maintenance.

There are disadvantages involved in this simplest of operations. It should be limited to applications where reproducibility and strength requirements are minimal. Since cleaning will also be minimal, the adhesive should be worked into the surface whenever possible to help remove and emulsify any remaining harmful surface film.

"TWO-PART PLUS" ADHESIVES

A step up from the one-part, minimum-process technique allows the use of two-part adhesives—primarily epoxies—in the low and intermediate cure-range covered in Chapter 4 as MIL-A-8623,

Types I and II. Many two-part epoxies, not formulated to meet any particular specification, may meet some particular processing need such as very fast cure for high production requirements, low viscosity for thin film application, and light-bodied thixotropic consistency for application as fillers or for application on vertical or overhead surfaces. They may be heavily filled with low expansion, metallic or inorganic fillers for metal-mending applications or for encapsulation of low-expansion components. Where weight saving is a requirement, the filler may be a low-density aggregate or microballoons, which are available as ceramic, phenolic, or glass.

In production the use of adhesives consisting of two or three parts requires careful weighing to insure reproducibility. Volumetric methods may be used for very fluid additives such as organic solvents or some of the reactive diluents commonly used with epoxy resins, but if any of the components are viscous enough to leave an appreciable film on the container, then the volumetric method becomes a source of error.

Acceptance Testing

Another concern is acceptance testing of the materials being mixed. If these are purchased according to a specification, then the responsibility rests with the supplier; if not purchased to a specification and the application is critical, some basic testing equipment and procedures are needed. A viscosity-measuring device as described in ASTM D1084-63 might be used. Solids content might be important in the case of solutions, and could be checked by one of the methods listed in ASTM. ASTM D-1338 could be used to determine working life, and ASTM D-2183 could be followed in evaluating the flow properties of dry-film adhesives.

Scales or balances are of prime concern and may vary in cost from less than $100 for simple, spring-loaded or beam-type scales to over $1,000 for automatic weight-balanced devices which feature accuracy along with rapid operation. Here some judgment will have to be made as to the frequency of weighing and economics of time consumed in weighing as well as accuracy needed.

OVENS

Even though an adhesive is normally considered to be room-temperature curing, improved bond-strengths are usually obtained with an oven cure. Table 6-1, listing the properties of twin-tube kits, shows that a 2-hour cure at 74C (165F) is sometimes better than a 5-day cure at room temperature. Also, oven curing saves considerable time.

Ovens are such useful gadgets that even a minimum setup should include one. Purchasing a gravity-convection type of oven can be a false economy. Adhesive cures are sensitive to temperature and time at temperature. In a gravity-convection type of oven, temperatures vary considerably from one part of the oven to another, and this may mean the difference between a well-cured and an uncured bond. For instance, a one-part, paste adhesive using a dicyandiamide curing agent may cure well at 149C (300F) for 3 hours, but cures tend to be marginal at 135C (275F), as is shown in Fig. 9-1, where 20 to 24 hours were required to achieve maximum strength. Dicyandiamide melts at 205C (401F). Cure is not initiated until it melts or dissolves and is in solution with the epoxy resin. Thus, if the oven temperature is near the solubility temperature of this particular curing agent, temperature control becomes important. Gravity-convection ovens can vary as much as 11C (20F), while a circulating-air oven can reasonably be expected to maintain ± 3C (± 5F) at temperatures up to 260C (500F).

ASTM Tentative Specifications for Gravity-Convection and Forced-Ventilation Ovens, designated E-145, sets standards for various types of ovens. These will vary from ± 2 percent and ± 5 percent for gravity-convection ovens to ± 1 percent and ± 2.5 percent for forced-ventilation ovens. The specification describes the method of testing for various oven variables.

Circulating-air ovens do require some attention with regard to overcrowding and spacing. If air flow is impeded, temperature variations can be considerable. Experience with a large industrial oven has shown that temperatures can vary by as much as 55C (100F) if overcrowding blocks off the free flow of air.

Glue-Line Temperature

One area of error which occasionally crops up during the cure of bonded assemblies—especially in ovens and in autoclaves where heat is conducted through air to the bond—is the interpretation of cure temperature. Cure temperature is always meant to be temperature at the bond line, or "glue-line temperature." The fact that an oven is preset at the desired temperature gives no indication of the temperature at the bond line. If the alignment and pressure tooling is massive it may take hours for the bond line to get to the

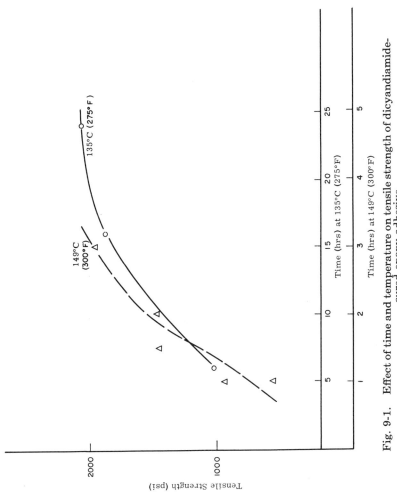

Fig. 9-1. Effect of time and temperature on tensile strength of dicyandiamide-cured epoxy adhesive.

oven temperature. It is a wise precaution to set thermocouples in or near the bond line at several locations for the first few runs. This will give a good estimate of the time necessary for the bond line to get to temperature. This time is then added to the cure time to arrive at an oven time necessary for a good cure. For some adhesives such as the nylon epoxies, where time to temperature is critical (usual requirement is less than one hour to temperature), this type of temperature monitoring could point out deficiencies in the heating unit and/or tooling design.

Pressure Limitations

Ovens are ideal for small parts and for curing many parts at one time. Pressures must be related to the type of adhesive used. Pressure applicators such as spring-loaded clamps should be adjustable to give reproducible pressures and so insure uniformity.

The use of conventional ovens with C-clamps for applying pressure makes it possible to use practically all presently available adhesives. Circulating-air ovens which achieve temperatures higher than 538C (1,000F) are available, so that polyimides and other newly developed high temperature adhesives which require cure temperatures as high as 371C (700F) can be processed. Springs used to apply pressures at these temperatures should be made from heat resistant alloys.

As parts to be bonded become larger, the use of various types of clamps becomes impractical and misleading, since clamps tend to localize the applied pressure. The obvious remedy of using heavy metal bars to distribute the clamping pressures or more clamps, placed closer together, soon becomes unwieldy. The next step is to go to large heated presses and/or autoclaves.

PRESSES

Presses vary in size from small laboratory equipment with 6-inch-square platens to industrial giants able to bond contoured assemblies as large as 6 feet by 18 feet. The platens of the large presses are heated either by steam or electricity. They should also be water-cooled to expedite cooling and prompt removal of a part, and insertion of the next assembly to be bonded; otherwise expensive equipment is tied up during a long cooling cycle.

Flat Bed Presses

The usual flat-bed press applies pressure to the part directly with the platens, through hydraulically activated rams, sometimes with two or more rams being used in one press. The principal problem here is to insure uniform distribution of pressure to the part being bonded. This is especially critical with contoured parts using contoured tooling. DeBruyne describes the use of thin, wood veneers over the parts being bonded.[2] These would compress in areas of high, localized pressures thus allowing distribution of pressure to other areas. Vulcanized-rubber pads are also used for this purpose; silicone rubber can be used at temperatures of 177C (350F) or higher. A more sophisticated method of distributing pressures evenly involves the use of unvulcanized rubber sheet over the part being bonded. As pressure is applied, the unvulcanized rubber flows to distribute pressure evenly and vulcanizes to maintain even pressures during the cure.

Cavity Press

The next stage in the evolution of presses is the cavity press, in which a concave, steam-heated water-cooled platen fits over a flat, steam-heated water-cooled platen with a rubber seal around the edge, as shown in Fig. 9-2.

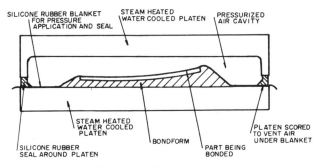

Fig. 9-2.   Cavity press.

The sketch in Fig. 9-2 also shows a contoured part being bonded. The part resting on a bond form with a matching contour is covered with a silicone-rubber blanket for pressure application and seal. Air, or inert-gas pressure is supplied to the cavity for application on the part being bonded. Cavity maximum operating-temperatures and pressures are usually 177C (350F) and 175 psi.

The limiting factor in the use of cavity presses is the size and contour depth of the part. A cavity press with a depth of 12 inches can accommodate a part with a contour depth of 10 inches. Figure 9-3 shows a view of a large cavity press in the open position. Cavity

Fig. 9-3.   Large cavity press in the open position.

presses can be as large as 18 feet by 6 feet with a cavity depth of 12 inches. Autoclaves are the answer to the need for assembly and curing of parts too large for cavity presses.

AUTOCLAVES

As aircraft panels became larger and contours more complicated, the available presses were finally strained beyond their capabilities. Autoclaves then evolved as the answer to the problem.

Vacuum Pressure

Where atmospheric pressure is sufficient to hold an assembly together during cure, it is possible to do without either a press or an autoclave. This type of assembly is usually done with flat panels or panels of simple contours where light pressures suffice to hold mating surfaces in contact. Such an assembly can be laid on a steam or electrically-heated, water-cooled table. A rubber blanket or plastic sheet such as polyvinyl alcohol is placed over the assembly and sealed to the table with a bead-type sealant. The table is provided with a pipe or hose outlet to a vacuum pump. When the vacuum is applied, atmospheric pressure holds the assembly together. The table is then heated and the adhesive is cured. This type of equipment is considerably less cumbersome than a press but, as already mentioned, its use is limited to assemblies which need no more than atmospheric pressure.

This process can be varied in several ways. Heat can be applied by including a heating pad in the assembly where feasible. Where large panels are involved, the assembly may be laid up on a wheeled platform, and, with vacuum lines trailing, may be transported into an adjoining oven for the curing cycle. Vacuum pressure monitoring, with gauges placed at strategic locations, is advisable, since if a leak develops anywhere in the seal of the rubber blanket or plastic sheet, the whole operation may be nullified.

Autoclaves

The final step in the evolution to the autoclave took place when the assembly, instead of being placed in an oven, was placed in a heated pressure-chamber with vacuum outlets. As airplanes and other lightweight bonded structures became larger and more complex, autoclaves achieved enormous proportions.

Some idea of the range in autoclave sizes is seen in Fig. 9-4; man is literally dwarfed by comparison. Figure 9-5 shows a rack of bagged, contoured panels being maneuvered into an autoclave. Figure 9-6 illustrates the use of a lift as a loading aid, combined with rails to guide the loading rack into the autoclave.

Autoclaves can be heated with steam coils; however, high-pressure steam lines present a limiting factor because of the danger of live steam escaping should the pipes break down. For temperatures up to 204C (400F), circulating hot-air is used to heat the autoclaves. To minimize the possibility of fire, inert gases such as carbon dioxide

Fig. 9-4.   Autoclaves at the Riverside plant.

Fig. 9-5.   Bonded cargo petal doors for the Lockheed C-141 Starlifter coming out of one of the autoclaves.

or nitrogen may be used. Most autoclaves are designed for a maximum pressure of 200 psi. Cold air is circulated through the autoclaves after the cure to drop the temperature as quickly as possible.

Fig. 9-6.   Lift and rail combination for servicing autoclave.

With hot-air or inert-gas heating, temperature-rise times are about 2C (36F) per minute. However, with some adhesives such as nylon epoxies, this is too slow. In order to increase the heat-up rate to 3C to 6C (37F to 43F) per minute, electrically heated platens are used in the autoclave to augment the hot-air heat.

Tooling design is an important factor in autoclave heating and cooling. Tooling should be designed to be as light as possible and to allow for maximum air-circulation.

TEMPERATURE

Once the decision is made to step up from room-temperature to elevated-temperature cures, temperature and its measurement are all-important. Massive tooling should be avoided because, acting as a heat sink, it can govern temperature distribution in the part being bonded, simply by slowing down the temperature-rise of parts in contact with it. If the use of massive tooling is unavoidable then it is necessary to monitor closely the temperature in the vicinity where the tooling contacts the adhesive bonds being cured, since this is the worst condition, and temperature-rise time is recorded for worst-condition locations. This practice should be followed whether the application is a simple assembly jig in a small oven or a large aircraft panel in a giant autoclave. Temperatures can be recorded by visual observation of thermometers inserted at strategic points in the parts being bonded. However, this is often difficult to do and may necessitate frequent opening of an oven thereby interfering with uniform heating. The use of thermocouples attached to recording or manually-operated potentiometers is a much more efficient solution to the problem of temperature monitoring. Thermocouples are small enough to be located exactly at the various "worst condition" bonding areas and attached to multitrack recording potentiometers to give a complete time-temperature history of the part.

Since temperature-rise time is very important with some adhesive types, and is important in the economic sense that fast oven-turnover increases production, it may be found desirable to accelerate the heating-up of an assembly being bonded by initially setting the oven temperature higher than the part cure-temperature; then, when the part has reached the cure temperature, the oven temperature may be allowed to drop back to the cure temperature.

Calibration of all temperature-monitoring equipment should be done at regular intervals and not be left to chance.

PRESSURE MONITORING

Pressure, regardless of how it is applied, should be known; otherwise one may not have a reproducible bond or, more important, a reproducible product. Pressure by dead-weight loading is definitely reproducible, since the dead weight is constant and independent of the amount of adhesive squeezed out of the bond. Pressure applied by properly calibrated spring-loaded clamps is reasonably reproducible as long as only a small amount of adhesive is squeezed out. C-clamps are probably the least satisfactory pressure devices for achieving reproducible bonds, since one tends to tighten them excessively, so that highly-localized stresses may buckle the plates distributing the load. The glue line may then vary from very thin to very thick. In the case of film adhesives particularly, if there is excessive squeeze-out as the bond warms up, the C-clamp may lose pressure entirely. C-clamps should be used with caution; the load should be applied through rubber pads to provide some follow-up pressure as the adhesive squeezes out.

In vacuum bag cures, pressure gauges or pressure probes are used to monitor the pressure at different sections of the bagged item. This is especially important in large panels since pressure could be lost at one end and not be readily monitored at the other end of an assembly. Here again, continuous monitoring of both the autoclave pressure and bag pressure is important and is a necessary part of the history of a bonded assembly.

VACUUM VENTING

When a part is assembled prior to curing in an autoclave, it is bagged; then a vacuum is pulled on the bagged assembly to hold everything in place. The part is then placed in the autoclave, where heat and pressure are built up to the required levels. Early in my experience with autoclave bonding, the practice was to maintain the bag vacuum throughout the curing cycle. Eventually—accidentally or otherwise—the bag vacuum was vented during the cure. The adhesive being used was a nylon-epoxy film, and the process-control specimens were climbing-drum peel, aluminum-honeycomb panels per ASTM D-1781. The resulting climbing-drum peel values showed an increase in strength of the vacuum-vented specimens by a factor of almost 2, over the nonvented specimens. The results of the investigation are shown in Table 9-1. The vacuum-vented specimens averaged 48.3 pli while the nonvented specimens averaged 25.8 pli. Both the vented and nonvented specimens failed cohesively in the adhesive. Subsequently, all vacuum bags were vented to the atmosphere once the autoclave pressure was built up to the required level. Inquiries made of other plants using autoclaves for producing bonded assemblies showed that this was common practice. If we had inquired earlier we might have had fewer rejects.

While there was considerable speculation as to the reason for this improvement, microscopic observation of the adhesive layer indicated that the adhesive in the nonvented specimens was slightly sponged, as compared with the adhesive in the vented specimens. Thus, it seemed reasonable that the adhesive in the vented speci-

Table 9-1. Effect of Venting Vacuum Lines on Bonding Aluminum, Honeycomb-Sandwich Construction[a]—Climbing-drum Peel Values

Panels Cured with Vacuum Lines Vented, in. lb/in.			Panels Cured under Vacuum, in. lb/in.
Spec No	Spec No	Spec No	Spec No
1. 51.5	10. 48.5	19. 41.5	1. 24.0
2. 55.0	11. 51.5	20. 41.5	2. 21.5
3. 48.5	12. 47.0	21. 43.5	3. 20.5
4. 53.5	13. 40.5	22. 47.0	4. 23.0
5. 45.0	14. 44.5	23. 49.0	5. 25.0
6. 50.5	15. 48.0	24. 45.5	6. 26.5
7. 59.5	16. 45.5	25. 46.0	7. 28.0
8. 52.5	17. 50.0	26. 49.5	8. 32.0
9. 60.0	18. 47.0	27. 51.0	9. 32.0
Average = 48.6			Average = 25.8
Average Dev = 3.7			Average Dev = 3.3

[a] Climbing-drum specimens bonded with nylon-epoxy film adhesive.

mens should be cohesively stronger than the adhesive in the non-vented specimens. Venting the vacuum lines then would minimize the effect of volatiles which would tend to sponge the adhesive.

BONDING OF EXTRA-LARGE PARTS

Lower-Temperature Cures

Inevitably, as airplanes and other lightweight, bonded assemblies became larger and larger, the units became too large for autoclave processing. The answer to this was the development of film adhesives which would cure at temperatures lower than 121C (250F). One of these, the nitrile epoxy, has been described in Table 5-11. This type has been cured successfully at temperatures as low as 82C (180F) for 6 hours. A room-temperature-curing epoxy is described by Hilton.[3] These adhesives make possible the assembly of structures too large for autoclaves. Such assemblies can be vacuum bagged and cured in large, heated rooms or ovens, or can be assembled with both adhesives and rivets and cured at room temperature or in a large, heated room or oven.

An added advantage of using adhesives curing at 93C (200F) or less, is that there would be less danger of warping in large structures. Bonding of aluminum structures may be accomplished at temperatures well below those which might weaken the metal. These adhesives have the disadvantage of requiring refrigerated storage.

Film adhesives are being developed which include an internal, electrical-resistance heat source.[4] This means that conventional film adhesives could be cured under ambient conditions, since the heat could be introduced directly into the glue line by bringing the electric power generators to the construction site and connecting to the resistance heaters contained in the adhesive. This type of processing is still in the development stage and, to my knowledge, has not been used in a commercially usable structure.

INSPECTION OF BONDED PARTS

Industrial processing of adhesives has come a long way from the hot-glue pot of the cabinetmaker and will undoubtedly continue making progress toward the goal where structural-metal adhesives will be used in on-the-site construction just as easily as riveting and welding are used today. However, this day may have to wait until a foolproof inspection method is available for on-site use.

As mentioned before, one of the disadvantages of bonding as an assembly method is that a bond area cannot be inspected visually. Inspection is done by two methods, destructive and nondestructive.

Destructive Inspection

Destructive inspection may be carried out on process-control test specimens prepared from the same adherend and adhesive materials as the production parts. The process-control specimen, as the name implies, accompanies the production parts throughout the cleaning, assembly, and cure operations. The adhesives and adherends are all assembled at the same time and cured in the same press or autoclave. As an additional control, each part may be designed with an expendable tab as an integral part of the assembly. After the cure, the tab is removed and subjected to the same tests as the control test-specimens. Results are then checked against specification requirements. The part is accepted or rejected on the basis of the control-specimen results.

The rejected parts may subsequently be inspected nondestructively for final acceptance or rejection. Final rejection would result in systematic destruction to learn how good or bad the parts really were.

In initial production of critical parts such as primary, bonded structures for aircraft, where human lives are dependent on reliability, a sampling and destructive analysis of actual production parts may be included in the test program.

Nondestructive Testing

A nondestructive test is one which can be made on a finished, bonded production part without impairing the effectiveness of the part for its intended application.

Load Testing

If a bonded part has failing stress of 2,000 psi, but will only see a cyclic service-load of 200 psi, a program of testing could be carried out with a combination of (a) destructive testing with gradually increasing cyclic loads to establish a safety factor, and (b) nondestructive testing with cyclic loading at representative service loads to establish a reliability factor. The program could start from 100 percent testing in its initial phases, to reduced sampling and testing in the later phases of production as reliability assurance was confirmed.

Portashear Testing

The Portashear test, as developed by General Dynamics, is essentially a destructive testing technique limited to areas which would eventually be eliminated in the finished part, such as rivet holes, etc. This method is used in metal-to-metal bonds where relatively thin-gauge-metal is bonded to a heavy-metal piece. A tool, similar to a hole saw, cuts through the surface skin, isolating a disc of bonded-metal skin. The disc is then sheared off with a known load so that the ultimate bond-strength of the bonded disc is measured. Hopefully, this value is representative of the bond strength in the part. The area tested is then drilled out as it would be in normal production. Full details of this procedure can be obtained from General Dynamics, at the Fort Worth Division.[5]

Sonic Tests

Probably the simplest nondestructive test method for bonded joints is the tapping test. The tapping tool may be a coin, a hammer, or other hand-held object. Since hand-actuated tools may have a nonreproducible force, a more sophisticated tool would be a mechanically-actuated tapping device. The pitch of the audible sound is the measure of the bond quality. When one considers the different sounds that result from tapping different parts of a complex bonded structure, it is surprising that the method can be used with any success at all. However, sonic tapping as an inspection method is probably one of the oldest in history. The inspector tapping a bonded joint is only following in the footsteps of the little old winemaker tapping a barrel to check the wine-level or to check the soundness of the barrel staves.

Ultrasonic Testing

The ultrasonic method is basically the same as the sonic method, in that a sound wave is sent through the structure and reflected or attenuated. The equipment, however, differs. For ultrasonic testing, a crystal generating a high-frequency signal and an oscilloscope are used. One of the first ultrasonic testers was the STUB meter (STUB standing for the Stanford Ultrasonic Bond meters). This tester evolved into numerous proprietary devices,[6] all of which make use of one of the following techniques:[7]

1. *Pulse Echo Reflection.* Short bursts of ultrasonic energy are sent into the part. A well-bonded part will reflect small echoes from each interface. An unbonded interface will reflect a large echo since most of the energy will be reflected, rather than transmitted.
2. *Pulsed-Through Transmission.* Here the sound is sent through the part and picked up on the opposite side. A well-bonded part will transmit a large part of the signal. A badly bonded area will attenuate the signal.
3. *Sweep Frequency Resonance.* Here the incident energy is varied in frequency so that the resonant frequency of the assembly can be observed. This method is said to actually give a measure of bond strength.[8]

A proposed ASTM method entitled "Standard Practice for Ultrasonic Resonance Testing of Adhesive Bonds" is based on the third technique described above. Details of this method can be found in Part 16 of the 1968 edition of the Book of Standards, where it has been published for information.

Ultrasonic test methods in general require standard void-specimens consisting of bonded assemblies representing the type of assembly being inspected, that is, metal-to-metal, metal-to-honeycomb, metal or glass-resin laminates, etc. Known voids are built into these standards and the ultrasonic tester is then calibrated by first checking these standard specimens.

Thermal Transmission Methods

Other methods based on the thermal transmission properties of the bond have been and are being developed. A well-bonded panel will transmit heat uniformly while a flawed panel will transmit less heat in the area of the flaw. The problem is to develop a foolproof method of detecting nonuniform transmittal of heat. One method is to freeze the panel so that it develops a uniform frost-cover. The panel is then heated uniformly on the side opposite the surface being inspected, so that the thaw pattern indicates thermal transmission and consequently, the degree of bond uniformity.

Another method makes use of chromophors or temperature-sensitive dyes to depict the thermal transmissibility of the panel.[9, 10] In this method heat can be either transmitted or reflected.

Thermal infrared inspection with radiometric detection of thermal emission is also used in bond-flaw detection.[11]

REFERENCES

1. American Welding Society. *Welding Handbook*. New York: American Welding Society, 1964.
2. DeBruyne, N. A. *Opening a New Era in Aircraft Engineering*. New York: Ciba Co., Inc., 1953.
3. Hilton, R. J. "Room Temperature Curing Adhesive Film." In *Symposium on Structural Adhesives Bonding*, edited by Michael J. Bodnar. New York: Interscience Publishers, 1966.
4. Bandaruk, W. "Adhesive Bonding with an Internal Electrical-Resistance Heat Source." In *Symposium on Structural Adhesives Bonding*, edited by Michael J. Bodnar. New York: Interscience Publishers, 1966.
5. Herndon, C. F. "The Use of a Portable Shear Tester for Nondestructive Testing of Adhesive Bonded Joints," Internal paper for Convair, General Dynamics, Fort Worth Division, 1960.
6. Fokker Bond Tester   Shur-Lok Bonded Structures Lim., Santa Ana, California
   Coinda Scope   Alma York, Fort Worth, Texas
   NAA Sonic Test System   North American Aviation, Inc., Los Angeles Division
   Sonoray   Branson Instruments, Inc., Stanford, California
   Reflectoscope   Sperry Products, Inc. Division of Automation Industries, Inc., Danbury, Connecticut
7. Babb, H. E. "Ultrasonic Inspection." In *Symposium on Structural Adhesives Bonding*, edited by Michael J. Bodnar. New York: Interscience Publishers, 1966.

8. Smith, D. F., and Cagle, C. V. "A Quality-Control System for Adhesive Bonding Utilizing Ultrasonic Testing." In *Symposium on Structural Adhesives Bonding*, edited by Michael J. Bodnar. New York: Interscience Publishers, 1966.
9. Vari Light Corp., *Method Using Liquid Crystals*. Cincinnati, Ohio: Vari Light Corp.
10. United States Radium Corp. *Thermographic Phosphors and Contact Thermography*. Bulletin 40.40. Morristown, New Jersey.
11. Vettito, P. R. "A Thermal Infrared Inspection Technique for Bond Flaw Detection." In *Applied Polymer Symposia on Structural Adhesive Bonding*, edited by Michael J. Bodnar. New York: Interscience Publishers, 1966.

CHAPTER 10

Industrial Applications

AIRCRAFT INDUSTRY

The aircraft industry spearheaded the development and application of structural-metal adhesives and is the principal customer for these materials. An advertisement by a company specializing in the fabrication of aluminum honeycomb states that one Boeing 747 and one Lockheed C-5A together would use almost two acres of honeycomb.[1] This implies that four acres of adhesive film are used with the honeycomb.

Boeing 747

One Boeing 747 (Fig. 10-1) will use about 40,000 square feet of adhesive film, about 950 pounds of polysulfide rubber sealant, and about 50 pounds of silicone rubber sealant. The use of adhesives in aircraft has increased steadily with improvement in materials and processing, and will probably continue to increase except in those areas where limitations are imposed by the high skin temperatures of supersonic aircraft.

B-58 (Hustler)

As was mentioned, the B-58 Hustler was the most thoroughly-bonded airplane of its time, with about 4,500 square feet of bonded paneling per airplane.[2] The wings, which doubled as aerodynamic, load-bearing primary structures and as fuel tanks, were almost entirely bonded. Their structure included edge metal-to-metal bonds of nitrile-rubber phenolic adhesives with high peel, high temperature characteristics, and aluminum skin-to-honeycomb bonds with epoxy-phenolic adhesives. Both aluminum and fiberglass-resin honeycomb were used. Shapes varied from flat panels to highly contoured leading edges. The complexity of the contoured members required considerable ingenuity in press platen design to insure that skin-to-honeycomb contact was achieved throughout the bonded area. Considerable use was made of nondestructive testing tech-

Fig. 10-1.   Boeing 747.

niques together with destructive evaluation to insure that the panels were thoroughly bonded.

The development of the two-phase nitrile-rubber phenolic/epoxy film adhesive, and later, the nylon epoxy and nitrile-rubber epoxy film adhesives (see Chapter 5) helped to overcome some of the limitations of the nitrile-rubber phenolic film by allowing the use of one adhesive for both edge-bonding and skin-to-honeycomb bonding.

Other aircraft-bonding applications include glass-fabric laminations, core splicing, and core-to-edge bonding. These are usually done with an epoxy-phenolic paste adhesive, since its foaming capability enables it to fill in voids and contact large areas of honeycomb edges.

Primers compatible with the intended adhesive are commonly used on freshly-cleaned surfaces both to help preserve the surfaces and to facilitate subsequent bonding operations and minimize the effect of storage.

Commercial jet planes use adhesive bonding in horizontal and vertical tail surfaces; wing panels, ailerons, and flags; tail-cone assemblies; and nose radomes.

Fatigue Factor

As jets were adapted for shorter flights, the landing cycles for a given flight-time increased so that fatigue cycles for highly-stressed areas increased. To counteract the effect of increased cycling, bonding was combined with riveting to increase fatigue-resistance of the rivet joints (see discussion of fatigue in Chapter 8).

A new development is the simultaneous curing of a resin-glass laminate skin and bonding of the skin to the honeycomb core (Fig. 10-2). Both resin glass and aluminum core are used, depending on the properties required of the panel. Practically all door panels and wheel-well panels are of bonded honeycomb-sandwich construction.

Fig. 10-2.   Panel construction in which resin glass laminate skin is cured at the same time it is bonded to the honeycomb core.

In general it can be said that adhesives are used where weight savings can be achieved by bonded sandwich construction. This includes exterior primary-load sections and interior paneling and floors. Weight savings are reflected in increased payload and/or range. For example, helicopter rotor blades are completely bonded and include outer-skin laminations and bonds to inner, lightweight core materials and to longitudinal spars.

The B-58 Hustler used 800 pounds of adhesives, the F-111 used approximately 1,000 pounds, and the Boeing 727 is estimated to be using about 5,000 pounds per airplane.[3] As planes become larger, adhesives more versatile, and weight saving and fatique require-ments more exacting, it can be expected that the percentage of bonded construction per aircraft will continue to increase.

Sealants

While adhesives are the "glamour" products, let us not forget the sealants. Of the synthetics, polysulfides probably have the longest history of use in airplanes (see Chapter 7). They are unequalled as fuel tank sealants; in addition, they are used in fairing applications and serve as edge sealants on honycomb panels and as sealants at pressurized joints. Silicone sealants are making a name for them-selves where heat resistance is required, as in ducts and near ex-haust outlets.

Satellite Launcher

Outer space missile and satellite construction is another area where high strength with minimum weight is a prime necessity. When the Air Force launched eight communications satellites into orbit in January 1967,[4] it was revealed that the holding and dis-pensing structure was a truss framework of lightweight tubular aluminum. Tubing was bonded to tubing with collar-type joints and nylon-epoxy film adhesive. A tack primer was used both to protect the etched aluminum surface and to hold the adhesive film in place. The joints were individually cured with clamp-type electrical heaters set to give the correct heat-up rate and cure temperature. With a slight stretch of the imagination, one can conceive of using the same technique to assemble larger and much more complicated bonded structures on site.

SURFACE TRANSPORTATION

Automobile Industry

The automobile industry has approached the use of structural bonding with caution. Weight saving is not as acute a problem in automobiles as it is in aircraft, and time needed for cleaning, assembly, and adhesive cures does not compare favorably with time needed for welding, riveting, and nut, bolt, and screw assembling. On the production line, time is a valuable commodity.

Structural Bonds—Some structural bonding is being done and more can be expected in time. The outstanding example of structural bonding on automobiles is the brake-shoe assembly, where friction lining is bonded to steel. Hundreds of millions of these have been successfully bonded since 1949. The adhesive is a modified phenolic, usually nitrile-rubber phenolic. The elimination of rivets and rivet holes simplifies construction, increases contact area, and allows wear right down to the steel shoe rather than to the rivet heads. Transmission bands or clutch facings are bonded in similar fashion, although adhesives are modified, usually by increasing the phenolic content, to withstand higher temperatures.

Because of their low modulus, the nitrile-rubber phenolic adhesives are useful for bonding windows into metal frames where these are small enough to be oven cured. Here the combination of stress relief and structural capability is important.

For large glass areas like front and rear windows, ambient-temperature curing materials such as polysulfide adhesive-sealants are used as structural bonds and seals. Polybutene sealants are also used for these applications.

Glass-resin laminated body sections of highly contoured sports car bodies are bonded, usually with polyester adhesives. Epoxy adhesives could also be used in this application but are more expensive. Rivets are used as holding fixtures during cure. Low production quantities and the absence of high-speed assembly lines make bonding more attractive than the use of mechanical fasteners in this type of application.

Lightweight aluminum body-sections in trucks are being bonded in increasing numbers with rubber-resin type adhesives. Neoprene contact-adhesives are found useful in the assembly of large truck trailers because of their quick-bonding properties.

A very important structural-bonding application already mentioned in the chapter on fatigue is the vulcanized-rubber-to-steel bond of motor mounts.

General Bonding and Sealing—A unique bonding application which is a vibration-damping device rather than a structural bond is the hood-stiffener assembly where a vinyl plastisol or neoprene-based adhesive is used. This material is applied to the oil-coated surface of the automobile hood. The stiffener is assembled to the adhesive pattern and edge-welded to hold it in place. The assembly is then put through the various cleaning and paint-dipping operations. The adhesive is finally cured during the paint drying and baking cycles. In contrast to the extreme cleaning, assembly, and cure conditions of the usual structural-bonding operation, this operation is designed to fit in with conveyer-line assembly with no costly delays. This was made possible by the formulation of an adhesive which could tolerate and bond through an oil film and could be heat-cured after a cleaning and paint-dipping cycle. Welding would have marred the hood surface and would have required a surface-refinishing job before painting.

Before painting, the welded drip-rail around the top of the car is sealed from corrosion at the weld and given a smooth finish with a mastic sealer such as a polysulfide or vinyl plastisol. Weld areas in general are possible sources of corrosion, and all are sealed off with asphalt-base sealers before painting.

Weatherstrip Bonding

Mastic-type rubber cement, either from reclaimed rubber or based on neoprene, is used for bonding sponge-rubber weatherstrip on the doors, doorways, and trunks, also under the hood. For assembling the weatherstrip, a wet adhesive may be partially dried to an aggressive tack, or it may be allowed to dry to a tack-free state and then reactivated with a suitable solvent immediately before assembly.

Processability is an important property for adhesives used on an automobile assembly line. The adhesive used to bond the hood stiffener should be sufficiently light-bodied to extrude easily but thixotropic so that it does not flow or sag during the cleaning and painting prior to curing. Solvent type, mastic or fluid rubber-cement formulations are modified according to the demands of assembly-line application; that is, roll coating, spraying, or extrusion, and according to the open time requirement between application and assembly and drying time.

Sound and vibration deadening is an important application in a product which must run as quietly as possible. Silencer pads are bonded in strategic locations with rubber-based adhesives.

Decorative Bonding

Another area where adhesives are taking over from rivet and screw fasteners is the bonding of nameplates and trim items. Elimination of rivet and screw holes not only saves the time and equipment necessary for the operation but also eliminates a source of corrosion. Rubber-resin types, such as the contact cements, are ideal for this operation since they do not require holding fixtures and tend to increase in strength, with time. Since nameplates and trim have to be installed after the paint job, compatibility between adhesive and paint must be established to insure that one does not degrade the other.

The ultimate in structural bonding for automobiles still awaits the final vote of confidence. Certain members in the transmission and power train which are presently welded or riveted could more easily be bonded. They have been bonded and evaluated in experimental units and have performed well. The adhesives used are epoxies and rubber-resin types. This application would have to take continuous cyclic loading under extreme atmospheric-exposure conditions, including exposure to automobile oils and fuels and road contaminants. The final vote of confidence will undoubtedly be cast when the combination of economics and feasibility force the issue.

Boats and Trains

In the future, transportation, in general, should make increasing use of structural bonding similar to that used in aircraft, chiefly because greater emphasis is being placed on high-speed transportation both over land and water. Aircraft companies are expanding activities into these areas. High speed inevitably places greater emphasis on aerodynamic surfaces and on weight saving. Primary and secondary structural-adhesive bonding can help out in both areas. Boeing, for instance, has completed and delivered to the Navy a high-speed, hydrofoil craft. Air-cushion boats are just beginning to make their presence felt in high-speed transportation over water; the British are using them to cross the Channel.

One of the more rugged applications for structural adhesives is the bonding of overlay materials on hydrofoils to protect them from the eroding action of water at speeds up to 90 knots.[5] A nylon-epoxy film-adhesive used for bonding metallic overlays on an HY150 steel hydrofoil performed excellently as it logged a total of 55 hours at 30 knots, 470 hours at 55 knots, and 201 hours at 90 knots. The bonded hydrofoil was submerged for five months in the course of the test. When one considers that nylon-epoxy adhesives tend to be sensitive to water, this is a truly remarkable performance.

The United States Department of Transportation is encouraging the development of high-speed trains in an effort to revive the passenger traffic in the railroad industry and relieve the crush of automobile traffic on the nation's highways. Adhesives will undoubtedly play a role in these changes.

GENERAL METAL BONDING

Today adhesive bonding of heavy metal castings is fairly common.[6] Bonding of simple parts, cast separately, is much less difficult than making intricate one-piece castings. Sandblasting or chemical etching is used to clean the surfaces to be bonded. Bearings are bonded into metal housings. Where the bearing material is of the same coefficient of expansion as the housing, assembly is simplified. However, during assembly of cylindrical pieces when the mating surfaces are coated and one is pushed into the other, adhesive tends to be scraped away and voids are formed. This type of problem can be minimized in several ways: (1) A beveled or scarfed joint may be used (see Fig. 4-12A). (2) The orifice or female member of the joint is heated to expand it, the male member is coated, and the two parts are assembled. If a heat-cure adhesive is used, the male part of the assembly expands on heating, pressure is applied to the bond, and positive squeeze-out with an adhesive-filled bond-area results. (3) Wherever possible, if two materials of different coefficients of expansion are bonded together with heat-curing, adhesive systems, the material with the higher coefficient of expansion is placed on the outside.

Temperature Variables

If a steel insert, such as a bearing, is bonded into an aluminum housing the assembly should be cured at an elevated temperature—as high or higher than the assembly will see in service—with the precautions already mentioned. The bond will then always be in compression because the aluminum has a higher coefficient of expansion than the steel. If the aluminum is the insert, then during the cold part of a temperature cycle, the aluminum will tend to pull away from the steel. The resultant stresses could be much greater than the bond strength.

Metal Cabinets

Metal assemblies such as cabinets, telephone booths, and light standards are being bonded with both one-part and two-part epoxy

Fig. 10-3.   Prototype adhesively-bonded light pole in New York City.

adhesives.[7] While one-part adhesives are more durable in terms of heat and solvent resistance, size, as in the case of the light standards, limits the application to two-part ambient-cure materials (see Figs. 10-3 and 10-4).

Epoxy adhesives, unless formulated to be electrically conductive by the addition of conductive fillers such as silver powder, specially-treated copper, acetylene black, or powdered graphite are good insulators or dielectrics when cured. However, in the bonding of metal cabinets which require grounding, the epoxy adhesives are sufficiently fluid to allow metal-to-metal contact at every joint and thus achieve electrical continuity throughout the assembly without the use of electrically-conductive formulations.

Backup Bonding

In certain areas adhesives are in continual rivalry with soldering, welding, brazing, and the use of rivets and threaded fasteners. Quite often in a development program adhesives are evaluated as a backup to the aforementioned traditional fastening methods. On occasion, the use of these traditional methods may cause warping or may result in metal combinations which are sensitive to electrolytic corrosion. Glass-to-metal seals may crack because of temperature shock. The backup adhesive program may very well save the day and the product. Other shortcomings result if in one assembly a series of brazed and/or solder joints are required. This may call for a series of brazing or solder materials with successively lower melting points so that the preceding joints will not soften upon reheating. Adhesives avoid this difficulty, since most adhesive formulations can be reheated to their cure temperature, or higher, without any appreciable softening.

BONDING OF ABRASIVES AND CUTTING TOOLS

The abrasives industry uses adhesives, and doubtless will continue to do so. Phenolic-bonded grinding wheels have been in use for a long time. A one-part, epoxy paste with a dicyandiamide curing agent per MIL-A-8623, Type III, is used to bond abrasive powders or granules to a solid or hollow-metal shaft. The shaft is first warmed and coated on the end with the paste. The coated end is dipped into the abrasive dust or granules to pick up an even coating of the particulate matter. This combination is cured for 1 hour at 177C (350F). A final coating of the adhesive paste is then

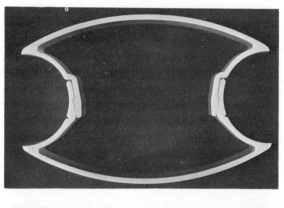

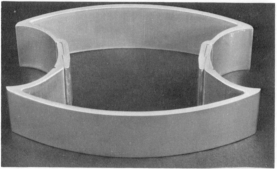

Fig. 10-4. Section of adhesively-bonded aluminum extrusions used in New York City prototype light pole.

put on the abrasive layer to embed the particles more firmly. This is then given a final cure of 1 hour at 177C (350F). In practice the epoxy wears away to expose the embedded abrasives. A heat-resistant epoxy formulation is necessary for this application.

Figure 10-5 shows a bonded cutting-tool consisting of a pure fused-alumina cutting-head bonded to a cold-rolled steel shaft with the same adhesive mentioned in the preceding paragraph. Here the very hard but brittle alumina is backed up by a tough steel holder. The adhesive makes the combination possible.[8]

IMPREGNATION OF POROUS CASTINGS

Cast-metal parts tend to be porous, and this must be remedied if they are to be used for such purposes as containing gases or liquids, sometimes under pressure. Vacuum impregnation of the porous

Fig. 10-5.   Fused aluminum cutting head bonded to cold rolled steel shaft.

metal with fluid epoxy formulations provides an easy solution to the problem, and is much less expensive than machining. Such a process, while not utilizing conventional sealants, is essentially a sealing rather than an adhesive type of application.

Since the pores may be filled with oily contaminants, thorough cleaning must precede impregnation. This may be accomplished by rinsing several times in a suitable solvent, if the pores are coarse enough, followed by oven drying.

Pore impregnation depends on the size of the porous casting and on the size of the pores. If the latter are large enough, immersion in the resin may be sufficient, since capillary action will suffice to fill the pores with resin. Small items with finer pores can be immersed in the impregnating resin contained in a vacuum chamber. A vacuum is pulled on the chamber to remove air from the casting; then atmospheric pressure imposed on the de-aerated, immersed casting will force the resin into the pores. An extension of this technique is to place the casting, immersed in resin, in a combination vacuum/ pressure chamber. These are commercially available items. De-aeration is then followed by pressure in excess of atmospheric pressure to force the resin into the pores.

The most extreme impregnation method with which I have been involved was used with a cast-aluminum container intended to hold a gas under thousands of pounds of pressure. Pressures of this magnitude revealed pores not likely to be sealed by the usual vacuum-pressure impregnation. Furthermore, the usual cleaning process did not flush out the finer pores. The process finally decided on was as follows:

1. The interior of the vessel was thoroughly rinsed in trichloro-ethylene. The trichloroethylene was forced through the pores of the casting by air pressure as high as 300 psi. The casting was then oven-dried at 71C (160F).
2. The vessel was filled with the impregnating epoxy resin warmed to 71C (160F). A pressure of 1,000 to 1,500 psi was used to force the resin into the pores.
3. Excess resin was poured from the impregnated casting. The casting was then placed in an oven at 121C (250F) to cure for 24 hours. When tested to destruction, this vessel failed by aluminum rupture at 5,500 psi.

The epoxy formulation used in this impregnation was resin A shown in Fig. 10-6. The data, as plotted, show that at 71C (160F) this formulation maintains a viscosity of about 0.5 poise for more than 2 hours, thereby allowing plenty of time for an extended impregnation procedure. The cure for this resin is 121C (250F) for 24 hours.

Formulation B has about 40 minutes of usable viscosity at 71C (160F) before the increase in viscosity becomes appreciable. The viscosity at ambient for formulation A is about 6 poises while for formulation B it is about 9 poises. The advantage of formulation B is that it has a considerably less stringent cure of 5 hours at 93C (200F).

These two are only a few of the many formulations which could be used for impregnation of porous castings. Viscosity, working life, and cure temperature are some of the more important parameters to be considered in making a choice. Another important factor is performance under the intended service conditions.

The important initial consideration is, given the use of good impregnating resins and techniques, are we taking full advantage of casting technology to make strong, impermeable, less-expensive pressure containers?

ELECTRICAL/ELECTRONIC INDUSTRY[9]

Bonded versus Unbonded Interfaces

In electrical or electronic applications, the various adhesives and sealants already described may be required to serve as sealants, encapsulants, conformal coatings, staking compounds, or impregnants. If electrical insulation is critical, then adhesion is a necessary re-

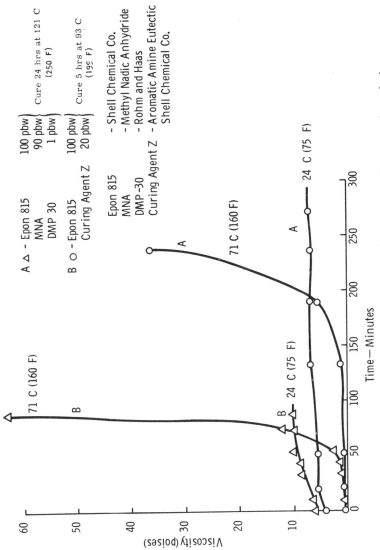

Fig. 10-6. Viscosity as a function of time and temperature for epoxy formulations.

quirement for all functions. The primary role of adhesives is usually to impart strength to an assembly. Sealants need adhesion to eliminate leakage paths for liquids and gases. In electrical/electronic applications, adhesion is needed to eliminate unbonded interfaces which may lead to electrical breakdown or even encourage short circuits, especially in high-voltage components and equipment. If components and printed circuits are well bonded, electrical properties of the equipment are dependent on the bulk electrical properties of the adhesive sealant or adherend. If, because of lack of adhesion, an unbonded interface is formed, then a path of low dielectric strength results. If a partial vacuum exists at the interface, the breakdown voltage becomes even less, in accordance with Paschen's law.[10] If moisture penetrates to the interface it may even become conductive.

Sometimes unbonded interfaces are inevitable, as in the case of glass vacuum-tubes or other glass-encased components protected by silicone-rubber envelopes and subsequently potted in epoxy resins. Successful bonds of nonsilicone adhesives and sealants to cured silicone-rubber are difficult, if not impossible to obtain. Electrical failures have been observed at unbonded interfaces resulting from such combinations. Such problems have been circumvented by using silicone-rubber sleeving over the leads. One end of the sleeving is embedded in the silicone-rubber sealant, protecting the component, while the sleeving itself provides insulation for the full length of the lead. The unbonded interface is no longer continuous and is interrupted at every lead.

In their early phases electrical and electronic applications, especially those intended for home consumption, used adhesive and sealant materials which were not expected to endure more than the ambient interior conditions found in a home or factory. These were materials such as pitch or wax for sealing capacitors, polystyrene coil dope, and cellulose-nitrate cements for bonding. Metal soaps such as zinc stearate were used for encapsulation. As industrial, military, and space applications developed, the range of temperature and environmental extremes of humidity, vibration, and radiation increased considerably.

The phenomenon of superconductivity at cryogenic temperatures has resulted in the development of supermagnets which make use of adhesives and sealants at liquid-helium temperatures.[11] Epoxies, polyurethane, and nylon epoxies have all been found useful at these temperatures[12] (see Fig. 5-8).

Printed Circuits

Printed circuits are made possible by the adhesive which bonds the copper to the circuit board.[13] For epoxy-glass boards, the adhesive may be vinyl phenolic, nitrile phenolic, or modified epoxy, depending on the required characteristics. Those mentioned would have a maximum-use-temperature of about 150C (302F). Glass-silicone laminated, printed circuit-boards would be used for temperatures as high as 260C (500F). Copper/Teflon® combinations also find use at temperatures as high as 260C.

Rigid, printed-circuit boards assembled in stacks can be bonded together with either thermoplastic or thermosetting adhesive-coated plastic film.[14] The adhesive-coated polyester, or other film, is alternated with the printed circuits in the stack. Heat, 125C to 155C (257F to 311F), depending on adhesive type, and pressure (about 10 psi) are then applied to bond the stack. For silicone and Teflon printed circuits, silicone-RTV adhesives have been used for similar assemblies. Teflon requires treatment with a sodium naphthalene complex (ASTM D-2093) and a silicone primer in preparation for bonding. The use of adhesive-coated dielectric films insures positive insulation as well as bond strength.

Flexible Printed Circuits

Complex assemblies sometimes require thin, flexible printed circuits (see Fig. 10-7). RTV silicone-rubber coatings are available not only for use as conformal insulating coatings but also to damp-out vibration which would put an undesirable fatigue-load on all solder joints, etc. Epoxy and silicone adhesives are also applied to component leads to strengthen them mechanically and ruggedize them against shock and vibration.

An important application (sometimes overlooked) for leads on components such as vacuum tubes, transistors, resistors, etc., is to put a kink or pigtail loop in the lead and then cover the lead with an elastomeric sealant (polysulfide or silicone) prior to potting in a rigid epoxy. The epoxy during cure, or during thermal cycling after cure, may impose tensile and compressive stresses on the component leads. The kink, or loop, encased in an elastomeric seal, gives easily to relieve and minimize such stresses, which, when unrelieved, might break component leads.

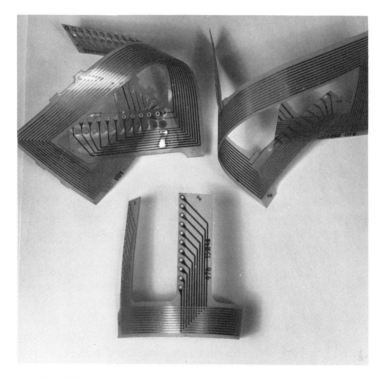

Fig. 10-7.   RTV silicone-coated flexible printed circuits.

Pressure Sensitive Components

Glass-encased diodes, transistors, vacuum tubes, etc., can be protected by coating with elastomeric resins such as polysulfides or silicone RTVs, preferably sponged or foamed to provide added extensibility and compressibility. Sponged silicone-rubber tubing can also be used over sensitive components. The ends are then easily sealed with the one-part silicone-rubber sealants. The most sophisticated approach to the problem of protecting expensive pressure-sensitive components is to mold internally-ribbed envelopes (see Fig. 10-8) from silicone or other suitable rubber. The open end is sealed after assembly. Toroidal transformer cores which are especially sensitive to pressure can be encased in very soft silicone sponge before being encapsulated.

Fig. 10-8. Vacuum tube protected by internally ribbed silicone rubber envelope.

Coils

Coils are present everywhere, in solenoids, magnets, and electrical assemblies in motors, etc. Adhesives and sealants are used to bond the coils to bobbins, to impregnate the windings, to bond the coils into holders, and also to hold down the leads to protect them against shock and vibration. Such measures upgrade performance considerably.

Coils can either be impregnated after winding, with fluid epoxy formulations, or they can be wet-wound, with the wire passing through the catalyzed epoxy on its way to the winding machine. Polyamide-cured epoxies have been successfully used as room-temperature-curing, wet-winding resins. These are relatively viscous and coat the wire well as it winds. Bonds to holders and for fastening leads may require thixotropic formulations as found in the QPL for specifications such as MIL-A-8623, Type I or II. If the coils are expected to operate at temperatures up to 74C (165F), then most room-temperature-cure hardeners such as polyamides or aliphatic polyamines should be satisfactory. At temperatures up to 177C (350F), aromatic amine hardeners such as metaphenylene diamine, methylene dianiline, or anhydride hardeners should be used. For higher temperatures up to 260C (500F), silicone resins should be considered. Coarse windings can be impregnated with the more fluid RTV-silicone rubbers. Otherwise, more conventional silicone impregnants must be used.

The use of silicones in electrical equipment, especially as impregnants for motor windings, has increased efficiency considerably by allowing continuous operation at higher temperatures than are usually considered normal.

A unique feature of silicones used in electrical equipment as sealants around connectors and leads is that under conditions of extreme heat, as in a fire, they do not become conductive and short out. Carbon-based resins such as epoxies, polyurethanes, etc., degrade to carbon, which is electrically conductive. Thus, electrical leads sealed with such resins would tend to short out in a fire. Silicones degrade to a nonconductive silica ash and would "fail safe" as an open circuit.

Resistance versus Temperature

The data in Table 10-1 give some idea of the drastic change in resistance which takes place as a carbon-base resin degrades when exposed to heat. The specimen used to obtain the data consisted of

Table 10-1. Resistance of Epoxy Resin* Exposed to 815C (1500F)

Time, min	Resistance, ohms	Time, min	Resistance, ohms
0	>5 × 10^6	9	0.12 × 10^6
1	>5 × 10^6	10	0.07 × 10^6
2	3 × 10^6	11	0.02 × 10^6
3	2 × 10^6	12	5000
4	1.5 × 10^6	13	1500
5	1.2 × 10^6	14	50
6	0.7 × 10^6	15	8
7	0.4 × 10^6	16	5
8	0.26 × 10^6	...	...

* Resin—Epon 828 (100 parts by weight).
Curing-agent Z (20 parts by weight) Cure 5 hours @ 121C (250F).

a ceramic ring with two copper leads wound around the ceramic as in Fig. 10-9. The ceramic ring was encapsulated in an epoxy resin and tested. The experiment was repeated with a silicone-RTV rubber encapsulant. The leads were connected to a DC ohmmeter and resistance was observed continuously after the encapsulated ring was placed in a muffle furnace preset at 815C (1500F). For the specimen with the epoxy encapsulant, the resistance decreased steadily from greater than 20 megohms (maximum reading on the ohmmeter) to about 5 ohms in 15 minutes, after which the experiment was discontinued. The silicone RTV under the same condition never read less than the maximum reading (20 megohms).

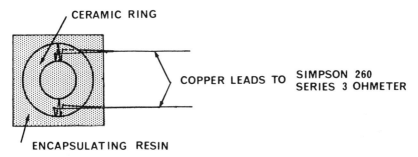

Fig. 10-9.   Specimen configuration for measuring resistance of encapsulant when exposed to 815°C (1500°F).

Miscellaneous

Laminated cores can be bonded after assembly by vacuum impregnation with fluid epoxy formulations as described in the section on impregnation of porous castings. An alternate method would be to coat the metal with the adhesive either before or after stamping out the individual parts. The adhesive in this case would be a solution of a B-stage epoxy or a vinyl-phenolic adhesive. The adhesive could be applied by spraying, dipping, or brushing. The coated parts as sheet metal could be dried for handling and storage. Subsequent assembly and cure with heat and pressure would then be a much cleaner and simpler operation.

Photomultiplier tubes sensitive to visible light sometimes use scintillator lenses to change invisible radiation to visible light. These lenses must be bonded to the tubes with adhesives which are transparent to the light wavelengths being transmitted. At the same time the adhesive must provide a stress-relief interlayer between the lens material and the glass shell of the tube. Water white silicone-RTV resins (Fig. 10-10), both one-part and two-part, have been used successfully for this application. Highly flexibilized transparent epoxies are also used for similar optical applications. Solar-cell panels are successfully sealed and coated (Fig. 10-11) as protection against the elements, including marine atmosphere and sea spray, with a water white RTV-silicone rubber. For best results a silane primer is used to improve adhesion of the RTV-silicone rubber. Less than 10 percent of the incident light is absorbed in this application.

Electrical components are quite often contained in metal or plastic containers, for which four types of seals are used:

1. *A structural seal of the cover to the can.* A rigid can-and-cover combination of like materials in terms of coefficient of expan-

Fig. 10-10.   Polymethylmethacrylate lens bonded to photomultiplier tube with water white RTV silicone adhesive.

Fig. 10-11.   Solar cells sealed and coated with a protective layer of water white room temperature vulcanized (RTV) silicone.

sion is successfully sealed with a rigid epoxy as specified in MIL-A-8623, Type II. Polyamide cured epoxies have also been used for these seals. If screws are used as the load-bearing fasteners, then epoxies or a polysulfide per MIL-S-8516 or a two-part RTV or noncorrosive one-part RTV-silicone could be used with the appropriate primer. A thin, flexible sheet metal or thermosetting plastic can-and-cover combination would be restricted to the elastomeric sealants described above since the rigid epoxies would not be able to take the peel or cleavage stresses resulting from the flexible adherends.

2. *Adhesive bonding of insulated conductors.* Feedthrough conductors with metal-to-glass seals may be used in the can or cover. These are sometimes soldered in place, but solder temperatures often tend to crack the seals. Adhesives, usually heavily filled with inorganic fillers such as alumina or silica, can successfully bond these insulated conductors in place. The curing agent, again, is dependent on the operational temperatures of the unit.

   Cylindrical preforms[15] made from dry one-part high-temperature-cure epoxies have been used for these seals. The advantage of the solid preforms is that they can be slipped into place around the conductor without danger of contaminating leads or contacts. The assembly is then heated to about 177C (350F) to soften and wet the mating surfaces and is subsequently cured. This operation would have to be done on the cover, or header, before assembly because of the temperatures involved.

3. *Sealing of electrical connectors with polysulfide or silicone sealants.* Electrical connectors are sometimes located on the can or cover. These are sealed either with a polysulfide per MIL-S-8516 or, if all electrical components are sealed, per MIL-S-8802. Where temperatures above 93C (200F) are to be encountered, silicone-RTV sealants should be used.

4. *Sealing of cables and wires with polysulfides or silicone-rubber sealants.* Sometimes instead of using feedthrough conductors or connectors, cables or wires are led directly through the container wall. Rubber grommets protect the cable insulation. Neoprene cables and grommets would be sealed with polysulfides. For temperatures higher than 121C (250F), silicone-insulated cable with silicone-rubber grommets should be sealed with silicone-rubber sealants. The noncorrosive one-part RTV sealants would be used.

If a resin sealant is used for the final seal in a container with an appreciable air-volume to be contained, care must be taken to cure the seal at room temperature. If curing is done at an elevated temperature, expanding air will cause blowholes to appear in the uncured sealant and a leaky seal will result.

Heat Sink Applications

In electrical/electronic assemblies the temperature rise due to evolution of heat from tubes, resistors, transformers, etc., in high-density circuits is quite often critical and a cause for concern. Design considerations must include thermally-conductive paths for removing heat from the circuitry involved. This circuitry may or may not be encapsulated. In confined circuitry as on a printed circuit board, not encapsulated, heat-sinks[16] bonded in place may be a sufficient answer. Aluminum is usually the preferred heat-sink material because of light weight and high thermal-conductivity. The adhesive bond should be air-free and be as thin as possible to minimize the heat-barrier effect. If good dielectric properties are required, the optimum combination of good electrical properties and good thermal conductivity can be achieved by using a high concentration of inorganic or mineral filler. This will increase thermal conductivity from 5×10^{-4} cal/sec/cm^2/1 deg C/cm for the unfilled epoxy to $20-30 \times 10^{-4}$ cal/sec/cm^2/1 deg C/cm for the filled epoxy.[17]

If electrical conductivity is not a deterrent, then a fine-mesh metal screen [18] or metal felt[19] impregnated with a clear epoxy adhesive forms a good heat-conducting adhesive. Sufficient pressure should be used on the impregnated metal screen or felt to insure that metal-to-metal contact is maintained during cure.

Heat-generating epoxy-encapsulated units would use the same guidelines for dissipating heat. Where good electrical properties are required, mineral-filled epoxies should be used. Where electrical conductivity is not harmful, metal screens or felts, strategically located, could help dissipate the heat.

Outgassing

Synthetic resins all have appreciable vapor pressures compared to inorganic materials such as metals and ceramics. The outgassing products causing the vapor pressure are objectionable because they may contaminate the atmosphere in an electronic assembly and encourage corona arcing; they may condense on electrical contacts resulting in increased contact resistance; or they could collect on dielectric surfaces and cause decreased surface resistivity. In un-

Table 10-2. Properties of Rigid and Semirigid Epoxy Resins

Formulation* Adhesive	Parts by Weight	Test Temperature Tensile Adhesion, psi Ambient	Test Temperature Tensile Adhesion, psi 177C (350F)	Elongation at Ambient Temp of Resin Alone Percent	During Cure Weight Loss, percent	After 24 hrs at 177C (350F) Weight Loss, percent
Epon** 828 Shell Curing-agent Z†	100 20	2162	1089	<5	−0.10	−0.28
Epon 828 Epon 871 Aminoethylpiperazine	40 60 15.5	3859	198	65	+1.00	−1.72
Epon 828 Epon 871 Shell Curing-agent Z†	20 80 12.3	2577	179	83	−0.12	−2.38
Epon 828 Epon 871 Diethylene Triamine	20 80 6.3	1109	163	41	−0.37	−2.38
Polysulfide Modified Epoxy Resin	...	3694	333	10.2	−0.18	−3.46

* All formulations cured for 16 hours @ 74C (165F).
** Epon — proprietary designation for Shell Chemical Co. epoxy resins.
† Aromatic amine curing-agent.

bonded interfaces outgassing may provide a low resistance electrical leakage path thereby encouraging an electrical short.

The bonded segments of the zero gradient Synchrotron[20] at Argonne National Laboratories present an outgassing problem in that continuous pumping is required to maintain a low enough vacuum so the high energy particles are not seriously impeded by the gas molecules while being accelerated in the Synchrotron.

Corrosion

Corrosivity of the resin and/or gaseous products may complicate the problems further. Copper and brass contacts may corrode. Gold plated, copper contacts may show corrosion if the gold plating is porous. Fine-wire coils or leads may corrode through, on long-time storage, causing open circuits.

For rigid, tightly crosslinked epoxy systems, outgassing seems to be limited to absorbed atmospheric gases such as water and carbon dioxide.[20, 21] Flexibilized epoxy systems (Table 10-2) and elastomeric adhesives such as nitrile-rubber phenolics[13, 20] tend to outgas more than the more rigid systems because of greater absorptive capacity and a lower order of polymerization.

Corrosivity has been associated with the more active room-temperature cure systems using polyamine curing-agents such as diethylene triamine and triethylene tetramine.[9] The corrosive potential of one-part silicone-rubber adhesives which have an acetic-acid volatile by-product has already been mentioned. Polysulfides may be corrosive, although polysulfide connector sealants meeting MIL-S-8516 have been found acceptable at temperatures up to 74C (165F).

With regard to outgassing, it is important to keep in mind that it does occur and can be corrosive. The problem, then, is to balance the convenience of processing against the sensitivity of the assembly to outgassing and corrosion. The problem is less severe for well-ventilated assemblies and for assemblies that are completely sealed from the outgassing products. The problem is greatest where the outgassing products are trapped in the same area with the sensitive components for extended periods.

REFERENCES

1. "Honeycomb: The Giant Jets Use It by the Acre." *Aviation Week and Space Technology* (January 8, 1968): 54-55.
2. Lunsford, L. R. "Design of Bonded Joints." In *Symposium on Adhesives*

for Structural Applications, edited by M. J. Bodnar. New York: Interscience Publishers, 1962.

3. Garrett, B. R. "The Growth of Adhesive Bonding." *Western Machinery and Steel World*, Vol. 58, No. 11 (1967): 29-32.

4. Hillesland, H. L. "Epoxy Adhesive Bonds Aluminum Satellite Dispenser," *SAMPE Journal* (April-May 1968): 63-65.

5. Lehman, A. F., and Trepel, W. B. "Evaluating Adhesives for Hydrofoils." *Materials Research and Standards* (September 1967); 383-389.

6. Wick, C. H. "All About Adhesive Bonding." *Machinery*, Parts 1-4 (October 1967-January 1968).

7. "The Pros and Cons of Adhesive Bonding," Staff Report, *Metal Progress*, May 1968.

8. "Adhesive Bonding Holds Down Costs," *Metal Progress*, May 1968.

9. DeLollis, N. J. "The Use of Adhesives and Sealants in Electronics." *IEEE Transactions on Parts, Materials, and Packaging*, Vol. PMP-1, No. 3 (December, 1965): 4-16.

10. Penning, F. M. *Electrical Discharges in Gases*. New York: Macmillan Company, 1957.

11. Caffey, H. T. et al. "A Protected 100 Kg Superconducting Magnet." *Journal of Applied Physics*, 36 (January 1965): 128-136.

12. Smith, B. M., and Susman, S. E. "Adhesives for Cryogenic Applications." Paper presented at the 1962 National Aerospace Engineering and Manufacturing Meeting, Los Angeles, California.

13. Rider, D. K. "Adhesives in Printed Circuit Applications." In *Symposium on Adhesives for Structural Applications*, edited by M. J. Bodnar. New York: Interscience Publishers, 1962.

14. G. T. Schjeldahl Co., Northfield, Minnesota.

15. Epoxy Products Co., New Haven, Connecticut.

16. Snogren, R. C. "Adhesives Bond Heat Sinks to Printed Circuit Boards," *Adhesives Age* (June 1968): 21-26.

17. Frados, Joel, ed. *Modern Plastics Encyclopedia*. New York: McGraw-Hill Book Company, Inc., 1966.

18. Newark Wire Cloth Company, Newark, New Jersey.

19. Huyck Corporation, Huyck Metals Department, P. O. Box 30, Milford, Connecticut.

20. Markley, R.; Roman, R.; and Vosecek, R. "Outgassing Data for Several Epoxy Resins and Rubbers for the Zero Gradient Synchrotron." *1961 Transactions*, Eighth Vacuum Symposium and Second Internation Congress, New York: Pergamon Press, 1962.

21. Baron, R. S., and Govier, R. P. *A Mass Spectrometric Study of the Outgassing of Some Elastomers and Plastics*. United Kingdom Atomic Energy Authority Research Groups, Rep. CLM-R16, 1962 (available from H. M. Stationery Office).

CHAPTER 11

Conductive Adhesives

Epoxy resins are noted for their good-to-excellent electrical properties. In the chapter on industrial applications, the electrical/electronic section described only those applications making use of the insulating or dielectric properties of the resin. This chapter, which is essentially a continuation of the chapter on applications, describes electrically-conductive resins, principally epoxies, and emphasizes the versatility and adaptability of these resins in meeting the requirements of industry.

Synthetic resins are made electrically conductive by the addition of either metallic fillers or conductive carbons. The carbon can be either an amorphous carbon such as acetylene black, or a finely divided graphite. Usually, finely divided silver flake is used in conductive epoxies and in conductive coatings. Silver works well because its salts and oxides are moderately conductive so that a slight amount of oxidation or tarnishing can be tolerated.

Non-noble metals such as aluminum or copper form nonconductive oxide coatings on each individual particle. Resins filled with these metals tend to be nonconductive. If the bond is formed under sufficient pressure to short through the metal-particle fillers, a misleading electrical conductivity results. A truly conductive resin should be electrically conductive in relatively thick bonds having no dependency on bonding pressure.

Resistivity or, inversely, conductivity is dependent on filler concentration. Figure 11-1 illustrates this factor.[1] The optimum concentration of silver is about 65 percent by weight.

Representative formulations are:

I. Epon 828*   100 pbw
   MD-750* (silver flake)   170 pbw   Cure—4 hrs @ 74C
   Diethylaminopropylamine   6 to 8 pbw   (165F)

This formulation would be used for intermediate temperature resistance with low outgassing requirements.

* The Metals Disintegrating Company supplied the MD-750, and the Shell Chemical Company supplied the Epon 828 and the Shell curing-agent Z.

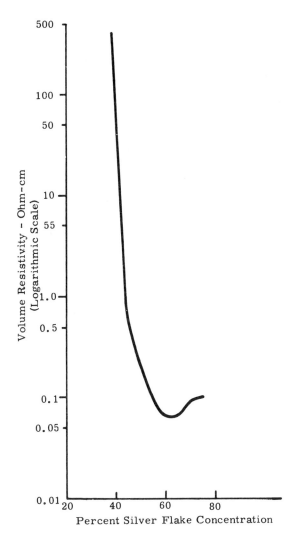

Fig. 11-1.   Variation in volume resistivity with silver flake concentration for conductive epoxy.

II. Epon 828*   100 pbw ⎫   Cure—5 hrs @ 93C
    MD-750* (silver flake)   175 pbw ⎬   (200F)
    Shell curing agent Z*   20 pbw ⎭

This formulation is for use in low outgassing applications with temperature resistance up to 149C (300F).

III. Epon 828*   35 pbw   ⎫   Cure—16 hrs @ 24C
      Polyamide curing agent   35 pbw   ⎬   (75F)
      MD-750* (silver flake)   130 pbw   ⎭   or
                                              2 hrs @ 74C
                                                 (165F)

This formulation would be used at temperatures up to 74C (165F). Wherever possible, cures should be at elevated temperatures since lower resistances are obtained with heat cures as shown in Table 11-1, where room temperature cures resulted in resistivities as high as 13 ohm/cm compared to a maximum of 0.048 ohm/cm with a 2-hour cure at 74C (165F).[2]

A one-part paste conductive epoxy could be prepared as in Table 11-1 with the exception that 4 parts by weight of dicyandiamide per 100 parts by weight of epoxy resin would be milled into the epoxy resin/silver mix. This should have a shelf life at room temperature of more than one year. Cure condition for a dicyandiamide-cured formulation would be 3 hours at 149C (300F).

INTERFACIAL RESISTANCE

Resistivity of a bonded specimen is very sensitive to the conduction of the adherend metal surface. Oxide formation could seriously increase the resistance of the assembly. Experiments (Table 11-2) using brass test plugs, with and without gold plating, show a decrease in resistivity by a factor of more than 10 when the test plugs are gold-plated.

Table 11-1. Resistivity of a Two-part Silver-loaded Epoxy Resin*

Material	Cure Time and Temperature		
	24 hrs @Room Temp	2 hrs @74C (165F)	4 hrs @74C (165F)
	Volume Resistivity, ohm/cm		
Silver-loaded epoxy with a polyamide curing-agent	1.028	0.0378	0.0135
	13.137	0.0134	0.0251
	4.064	0.0254	0.0279
	9.964	0.0477	0.0312
	0.490	0.0183	0.0417

Note: Specimens consisted of brass tensile plugs (1⅛ inches in diameter) bonded together with a 0.010- to 0.020-inch glue-line thickness.
*Taken from Reference 2.

Table 11-2. Comparison of Electrical and Physical Properties of Bonded Brass Plugs and Gold-Plated Brass Plugs*

	Kit #1		Kit #2	
	Brass Plugs	Gold-Plated Brass Plugs	Brass Plugs	Gold-Plated Brass Plugs
	Volume Resistivity, ohm/cm			
	0.0275 0.0314 0.0353	0.001270 0.001350 0.001456	0.0229 0.0192 0.0216	0.001700 0.001288 0.001524
Average	0.0314	0.001359	0.0212	0.001504
At 121C (250F)	0.0297 0.0331 0.0373	0.001318 0.001397 0.001185	0.0245 0.0209 0.0227	0.002149 0.001687 0.001758
Average	0.0334	0.001300	0.0227	0.001865
	Tensile Strength, psi			
At 121C (250F)	1540 1730 1800	2170 1410 1270	1170 1370 1380	1090 1220 1030
Average	1690	1617	1307	1113
Type of failure	Cohesive	Adhesive	Cohesive	Adhesive

* Adhesive formulation (approximate): epoxy resin (100 parts by weight)
silver flake (about 170 parts by weight)
dicyandiamide (4 parts by weight)
  Cure: 6 hours @ 135C (275F)

A similar experiment with the same conductive adhesive, comparing silver plating with gold plating, gave the results shown in Table 11-3. These data illustrate how easily a testing laboratory could unjustly penalize the conductive adhesive by improper preparation of the adherend surfaces.

Table 11-3. Comparison of Electrical and Physical Properties of Bonded Brass Plugs and Bonded Gold and Silver-Plated Plugs

Adherend Material	Bond Thickness, inch	Resistivity at Room Temp, ohm/cm	Tensile Adhesion at 121C (250F), psi	Type of Failure
Brass	0.012	0.015	2090	Cohesive
Silver-plated brass	0.010	0.0014	1440	Adhesive
Gold-plated brass	0.010	0.0015	1760	Partly cohesive

All values, average of five tests.

Mixing Time

Some work has been done to show that time taken to mix the silver into the resin can be an important variable, one-half hour being preferable to longer times such as 2 to 4 hours.[3] This, however, was a more important variable in the early days of conductive resins when each laboratory prepared its own mix. Today, more uniformity can be expected from commercially prepared formulations.

COPPER-FILLED EPOXIES

A good illustration of industry's efforts to improve the economics of a material whenever possible is the development of lower-priced conductive adhesives. Copper, having excellent electrical properties and costing much less than silver, was a good starting point. Efforts to improve the surface properties of copper powder were directed at first toward coating each particle with silver so that copper oxide or copper salts could not form. Thus, a silver-coated copper-powder filler became the basis for cheaper conductive epoxies. Such copper powders are effective, but results are not quite as good as with pure silver-powder fillers.

The next step was to incorporate ingredients in the resin formulation to protect the surface of the copper-powder filler and keep it conductive. Electrically conductive, copper-filled epoxy adhesives are now available which result in electrical resistivity comparable to that obtained with gold- and silver-plated brass plugs when bonded to unplated brass plugs. The values are shown in Table 11-4.

Table 11-4. Resistivity (ohm/cm) of Copper-filled Epoxy[a]

Resin Bond Thickness, inch	Initial	After One Week at 71C (160F)	After Two Weeks at 71C (160F)	After 24 hrs at 121C (250F)
		Volume Resistivity, ohm/cm		
0.010	0.0018	0.0035	0.0035	0.009
0.020	0.0013	0.0088	0.0025	0.012

[a] Since the formulation of the adhesive is proprietary,[4] its exact ingredients have not been revealed; it is known only that the resin contains about 82 percent copper powder by weight.

Since the initial resistivity values are comparable to those obtained for gold- and silver-plated surfaces, it is possible that the formulation tends to improve and protect the surface of the brass adherend. There is some indication that extended storage at elevated temperatures does increase the resistivity.

APPLICATIONS

In the chapter on applications, the use of conductive adhesives in heat-sink applications has already been described. Conductive resins are used as adhesives and coatings in microelectronic assemblies.[5] Such applications include attachment of fine lead wires to printed circuits, electroplating bases, metallization on ceramic substrates, grounding metal chassis, bonding wire leads to header pins, bonding components to plated-through holes on printed circuits, wave-guide tuning, and hole patching.

Conductive adhesives are used as substitutes for spot welding when welding temperatures build up excess resistance at the weld because of oxide formation.

Conductive adhesives form joints with sufficient strength so that they can be used as structural bonds where electrical continuity in addition to bond strength is required, as in shielded assemblies.

Conductive adhesives are used in ferroelectric devices to bond electrode terminals to the crystals and to bond ferroelectric crystals in stacks. Conductive adhesives replace solders and welds where crystals tend to be depoled by soldering and welding temperatures.

Battery terminals can be successfully bonded when soldering temperatures are harmful.

TEST METHOD

Most of the data presented so far in this chapter have been obtained by ASTM Method D2739-68T, Method of a Test to Determine Resistivity of Conductive Adhesives. The test specimen and recommended alignment jig are shown in Figs. 11-2 and 11-3. The method calls out cleaning and preparation procedures and test equipment. Since interfacial resistance may be a significant factor in total resistance, gold- or silver-plating is recommended where extreme accuracy is required.

ALTERNATE CONDUCTIVE SYSTEMS

Other conductive systems which do not compare with metal-filled systems, but which do have their uses, are resin combinations with

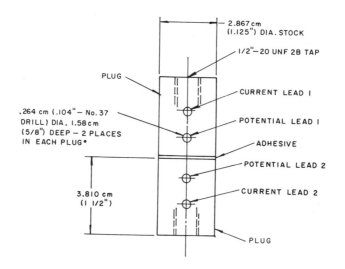

Fig. 11-2.   Brass tensile adhesion specimen with electrical connections.

carbon in the form of graphite or amorphous carbons such as acetylene black. Representative formulations and the respective resistivities are shown in Table 11-5.

Except for those evaluated in formulation Number 11 of Table 11-5, the specimens consisted of 2-inch-diameter discs with thicknesses as shown. They were coated with air-dried silver paint on both sides, prior to being evaluated. Measurements were made with a Leeds & Northrup Wheatstone bridge, a Rhode & Shwarz KMT holder, and a Kiethly vacuum-tube voltmeter.

Number 11 presents the average value for five specimens prepared with a carbon-filled polysulfide resin. These measurements of volume resistivity were made with a tensile-adhesion specimen similar to the ASTM 2739 specimen.

As is evident, the values are high with a considerable range in values. However, properly formulated, these materials can be used in applications such as the bonding of conductive floor tile in static-free rooms.

Silicone-RTV resins can also be filled with silver and carbon fillers to give conductive formulations. However, these are not yet in common use.

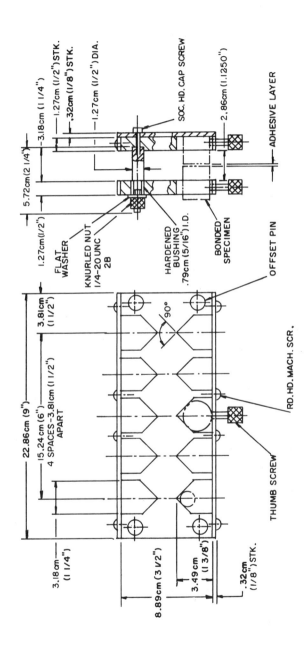

Fig. 11-3.　Sketch of suggested assembly jig (material—steel).

Table 11-5. Carbon-filled Conductive Formulations

Sample No.	Material and Parts by Weight*	Cure	Thickness, (cm)	Volume Resistivity, (ohm/cm)
1	Hysol 6020/graphite A/DEAPA (50/40/3)	3 hrs @ 74C (165F)	0.94	31,000
2	Hysol 6020/graphite A/DEAPA (50/35/3)	3 hrs @ 74C (165F)	0.94	3,000
3	Hysol 6020/graphite A/DEAPA (50/35/3)	Room temperature	0.39	43,000
4	Hysol 6020/graphite A/DEAPA (50/30/3)	3 hrs @ 74C (165F)	0.93	5,000
5	Hysol 6020/graphite A/DEAPA (50/30/3)	Room temperature	0.62	13,000
6	Adiprene L/acetylene black/MOCA (100/7.5/11)	16 hrs @ 74C (165F)	1.00	64,000

Table 11-5 (Continued). Carbon-filled Conductive Formulations

Sample No.	Material and Parts by Weight*	Cure	Thickness, (cm)	Volume Resistivity, (ohm/cm)
7	Adiprene L/acetylene black/MOCA (100/7.5/11)	Room temperature	1.00	180,000
8	Adiprene L/graphite A/MOCA (100/80/11)	Room temperature	1.02	53,000
9	Hysol 6020/acetylene black/DEAPA (100/10/6)	3 hrs @ 74C (165F)	0.95	110
10	Hysol 6020/acetylene black/DEAPA (100/10/6)	Room temperature	0.99	160
11†	Proprietary carbon-filled polysulfide sealant (3C 403HT)	Room temperature	...	340 (230 to 450)

* *Sources of Materials in Formulations:*
Hysol 6020 obtained from Hysol Corporation.
Graphite A = microfine Grade A obtained from Acheson Colloids Co.
DEAPA = diethylaminopropylamine.
Acetylene black obtained from Columbia Carbon Co.
Adiprene L = Adiprene L100 obtained from E. I. duPont de Nemours & Co., Inc.
MOCA = Methylene-bis-orthochloroaniline obtained from E. I. duPont de Nemours & Co., Inc.
† 3C 403HT supplied by Churchill Chemical Company, Los Angeles, Calif.

CONTINUOUS METAL CONDUCTIVITY

Up to this point, we have considered conductive formulations which are independent of pressure; that is, materials that are conductive under zero-pressure conditions in thick films. If pressure is allowed in an assembly to maintain contact across metal fillers during cure, it would be possible to prepare conductive adhesives with practically any metal filler either in particulate form or as a woven-metal screen or metal felt. The resistance across this type of assembly is sensitive to pressure and to type of filler or matrix, as is shown in Table 11-6.[6]

Table 11-6. Resistance Variations, using Gold-Plated and Nonplated Matrices

Matrices	Pressure Across Adherends			
	Brass (Sanded Only)		Gold Plated Brass	
	None	50 psi	None	50 psi
	Resistance, ohms $\times 10^{-6}$			
Tungsten, 140-mesh	10,000	797.0	8,700.0	492.0
Tungsten, gold-plated	2,070	60.8	682.0	22.8
Silver, 50-mesh	1,040	85.0	164.0	7.3
Silver, gold-plated	498	15.8	174.5	7.3
Copper, 40-mesh	1,100	168.0	760.0	28.0
Copper, gold-plated	461	18.6	237.8	11.5

Table 11-6 shows a decrease in resistivity by an overall factor of 10 or more for most combinations with a pressure increase to 50 psi. The use of gold-plated adherends and matrices results in a similar decrease in resistivity.

This type of conductive system would be successful when used between flat, evenly spaced surfaces to assure contact with the greatest possible number of points. Rough surfaces would limit contact to the few points of closest approach.

WIRE EMBEDMENTS IN CONDUCTIVE EPOXIES

One of the more widely-used applications mentioned for silver-filled epoxies is as a substitute for solder, brazing, and welding in bonding wires to printed-circuit terminals, fastening coil-terminations, bonding armature-leads to carbon brushes, etc. The resistance to be expected for such embedded-wire applications seems to have received very little attention in the literature; therefore, the following study of resistance of wire embedments in silver-filled epoxy adhesives has been included in this chapter.

Materials

The materials studied include three types of stranded wire: bronze, silvered copper, and tinned copper; three metal blocks: brass, aluminum, and stainless steel; and four silver-loaded epoxy resins: a dicyandiamide-cured epoxy, an aromatic-amine-eutectic-cured unmodified epoxy, a polyamide-cured semirigid epoxy, and a triethylene-tetramine-cured unmodified epoxy. The adhesive formulations are listed in Table 11-7.

The metal blocks were drilled with holes 3/32-inch in diameter by 3/16-inch deep. The Teflon® insulated wires were cut 2 inches long with the ends bared for ½ inch.

Table 11-7. Specifications of Adhesive Formulations

Designation	Materials	Parts by Weight	Cure
*For Tables 11-8 through 11-11:**			
IA	Unmodified epoxy	100	3 hrs @ 149C(300F)
	Aromatic amine-eutectic		
	hardener	20	
IIA	One part epoxy		
	dicyandiamide hardener	. . .	3 hrs @ 149C(300F)
IB	Unmodified epoxy	100	2 hrs @ 74C(165F)
	Triethylene tetramine	10	
IIB	Modified epoxy	50	2 hrs @ 74C(165F)
	Polyamide hardener	50	

*NOTE: Silver content for all the above formulations is approximately 65 percent.

For Table 11-12:			
A	Epon 828	100	3 hrs @ 149C(300F)
	Aromatic amine-eutectic		
	hardener	20	
B	Epon 828	100	2 hrs @ 74C(165F)
	Triethylene tetramine	10	
C	Epon 828	100	
	Aromatic amine-eutectic		
	hardener	20	3 hrs @ 149C(300F)
	Acetylene black	10	
D	Epon 828	100	2 hrs @ 74C(165F)
	Triethylene tetramine	10	
	Acetylene black	10	

Experimental

The metal blocks were cleaned by vapor-phase degreasing, sand-blasting, and ultrasonic cleaning in trichloroethylene and isopropyl alcohol, followed by a vacuum bake at 93C (200F). The wires were cleaned ultrasonically, as above, and vacuum baked. The holes were filled from the bottom up with the silver-loaded epoxy by means of a hypodermic needle. The wires were precoated with the silver-filled epoxy to ensure wetting, and were then bent into a U-shape and inserted into the holes. As the sets of the various epoxy-wire combinations were completed, they were put into the oven that was preset at the proper cure-temperature. After being allowed 10 to 15 minutes to warm up to oven temperature (during which time the silver-loaded epoxy became considerably more fluid), the plate with the wire-embedment was tapped sharply on a solid base to force air bubbles out of the assembly.

Measurement of resistance between the embedded wire and the metal block was made after the initial cure (see Table 11-8); again after 24, 48, and 72 hours at 127C (260F) for the A adhesives and 74C (165F) for the B adhesives; then, after 18 hours at −54C

Table 11-8. Initial Resistance of Embedded Wire[a]

Stranded Wire Type	Metal Plate					
	Brass		Aluminum		Stainless Steel	
	Adhesives[b]					
	IA	IIA	IA	IIA	IA	IIA
	Resistance, ohms					
Bronze	0.02	0.02	0.01	0.01	0.05	0.15
Tinned copper	0.02	0.02	0.52	0.06	0.12	0.08
Silvered copper	0.01	0.01	1.28	0.04	0.25	0.76
	Adhesives[b]					
	IB	IIB	IB	IIB	IB	IIB
	Resistance, ohms					
Bronze	0.01	0.01	0.01	0.01	0.03	0.03
Tinned copper	0.01	0.01	0.03	0.01	0.05	0.03
Silvered copper	0.01	0.01	0.05	0.02	0.12	0.02

[a] All values, average of five specimens.
[b] For formulations, see Table 11-7.

($-65F$) after a total of 166 hours at temperature; and finally, after a total of 332 hours at temperature. Table 11-9 gives the values after the complete exposure time. Table 11-10 gives the results for the low-temperature-cure adhesives after exposure to one week (168 hours) at 127C (260F) in addition to the previous exposure to 74C (165F). Table 11-11 gives the resistance of the same embedments as those shown in Tables 11-9 and 11-10 except that they have been potted in an epoxy resin.

Up to this point, all of the data have been obtained on one set of wires embedded in silver-filled epoxies, and the assumption is made that the silver filling has contributed appreciably to the results. When the wires were inserted in the resin-filled holes, they were bottomed out so that in every case metal-to-metal contact was probably achieved. Thus it seemed possible that the primary function

Table 11-9. Resistance of Embedded Wire[a]

Stranded Wire Type	After 332 Hours at 127C(260F) + 18 Hours at $-54C(-165F)$					
	Metal Plate					
	Brass		Aluminum		Stainless Steel	
	Adhesives[b]					
	IA	IIA	IA	IIA	IA	IIA
	Resistance, ohms					
Bronze	0.02	0.02	0.03	0.04	0.19	0.63
Tinned copper	0.04	0.03	> 6.49[c]	0.96	0.63	0.68
Silvered copper	0.03	0.05	>19.49	>19.56	>7.98	1.03

	After 332 Hours at 74C(165F) + 18 Hours at $-54C(-65F)$					
	Adhesives[b]					
	IB	IIB	IB	IIB	IB	IIB
	Resistance, ohms					
Bronze	0.02	0.02	0.02	0.03	0.02	0.03
Tinned copper	0.01	0.01	0.19	0.04	0.07	0.03
Silvered copper	0.01	0.01	0.17	0.01	0.14	0.04

[a] All values, average of five specimens.
[b] For formulations, see Table 11-7.
[c] Those averages with the "greater than" (>) sign have one or more individual values of >30 ohms (maximum for the ohmmeter used).

Table 11-10. Resistance of Embedded Wire[a]

After 332 Hours at 74C(165F) + 18 Hours
at −54C(−165F) + 168 Hours at 127C(260F)

Stranded Wire Type	Metal Plate					
	Brass		Aluminum		Stainless Steel	
	Adhesives[b]					
	IB	IIB	IB	IIB	IB	IIB
	Resistance, ohms					
Bronze	0.04	0.03	0.03	0.07	0.23	0.07
Tinned copper	0.15	0.01	0.08	0.17	0.23	0.12
Silvered copper	0.05	0.01	>15.00[c]	0.38	>12.17	0.19

[a] All values, average of five specimens.
[b] For formulations, see Table 11-7.
[c] Those averages with "greater than" (>) sign have one or more individual values of >30 ohms (maximum for the ohmmeter used).

Table 11-11. Resistance of Embedded Wire after Potting in Epoxy Resin[a] and after an Additional Exposure of One Week at 127C (260F)

Stranded Wire Type	Metal Plate[b]					
	Brass		Aluminum		Stainless Steel	
	Adhesives[c]					
	IA	IIA	IA	IIA	IA	IIA
	Resistance, ohms					
Bronze	0.09	0.06	0.11	0.19	0.35	1.07
Tinned copper	25.22[d]	0.04	34.9	14.13	5.41	0.43
Silvered copper	0.17	0.03	4,161.00	77.64	4.6	0.98
	Adhesives[c]					
	IB	IIB	IB	IIB	IB	IIB
	Resistance, ohms					
Bronze	0.06	0.06	0.08	0.08	0.35	0.19
Tinned copper	0.87	0.03	3.32	0.62	10.63	0.22
Silvered copper	1.03	0.02	1 to ∞	0.57	11.99	0.44

[a] Potting-resin formulation = epoxy resin, 100 parts by wt/hexahydrophthalic anhydride, 76 parts by wt/P 11-80 Dow Chemical Polyol, 66 parts by wt/ DMP-30 (amine), 2 parts by wt.
[b] All values, average of five specimens.
[c] For formulations, see Table 11-7.
[d] Includes one value of 125 ohms.

of the resin was to maintain the metal-to-metal contact which was the principal reason for the low-resistance values. This possibility was checked out in the final experiment, and the results are shown in Table 11-12, which gives the resistance values for wires embedded in unfilled epoxy resin and in carbon-filled epoxy resin before and after exposure to 127C (260F) for one week. Only brass and aluminum plates were used in these tests since these seemed to represent the extremes in the previous tests.

Table 11-12. Resistance of Wires Embedded in an Unfilled-Epoxy Resin and in a Carbon-Filled Epoxy Resin

Stranded Wire Type	Brass #1		Brass #2		Aluminum	
	Adhesive*					
	A	B	C	D	A	C
	Initial Resistance, ohms					
Bronze	0.78	0.4	<0.1	<0.1	0.1	0.3
Tinned copper	<0.1	∞	<0.1	55	0.1	2.02
Silvered copper	<0.1	∞	<0.1	<0.1	0.6	0.4
	Resistance after One Week at 74C(165F) plus One Week at 127C(260F)					
	Resistance, ohms					
Bronze	0.05 to ∞	0.04 to ∞	18.1	0.11	10.1	8.82
Tinned copper	0.06 to ∞	∞	45.8	60.3	0.09 to ∞	867
Silvered copper	0.05 to ∞	∞	0.86	21.0	184	56.4

* For formulations, see Table 11-7.

Data

Many more measurements were made than are shown in these tables; only those data are published which tend to summarize the results. Each value listed is the average of measurements made on five specimens. The first measurements were made on an ohmmeter which could be read to an accuracy of 0.01 ohm with a maximum capacity of 30 ohms. (This causes some confusion in Tables 11-9 and 11-10 because it is difficult to average five figures of which some are greater than 30 ohms.) By Table 11-11, the higher-resistance values were being measured on a Simpson ohmmeter which could be read sufficiently well up to 20 megohms with a minimum reading of 0.1 ohm. Infinity (∞) really means "greater than 20 megohms."

Only in one average in Table 11-10—listed as (>15)—is there a chance that infinity was reached without being observed. All other greater than (>) averages were finite values as measured for Table 11-11. In Tables 11-11 and 11-12 any group of five values containing individual measurements greater than 20 megohms are shown as a range of values including infinity.

Table 11-11 attempts to answer the question, What happens when embedded wires are vacuum-encapsulated in fluid unfilled-epoxy resin? The specimens used were those which had already been exposed as noted in Tables 11-9 and 11-10. With one possible exception the potting resin did not seem to have any significant effect. The instabilities all seem to be extensions of those already noted in Tables 11-9 and 11-10. The only possible exception in Table 11-11 is tinned copper in brass with adhesive IA, showing an average value of 25.22 ohms. While the one value of 125 ohms far outstrips the other four, those four do show some tendency to increase resistance, with an average resistance of 0.22.

Comparison of the data in Tables 11-9 and 11-10 indicates that after 332 hours of aging at temperature, the primary cause of degradation is metal-to-metal incompatibility, not silver-filled epoxy instability. The order of instability based on decreasing resistance is: silvered copper in aluminum, silvered copper in stainless steel, and tinned copper in aluminum. Tinned copper in stainless steel, and bronze in stainless steel show some instability but are still low, with a maximum resistance of 0.68 ohm. None of the wires in brass nor the bronze wires in aluminum show any significant change in resistance.

The combinations exposed to 74C (165F) show no change of resistance with time except for three values indicating incipient instability, with a high resistance of 0.19 ohm. Table 11-10 shows results for combinations containing epoxies cured with triethylene tetramine and polyamide after being exposed to 127C (260F) for an additional 168 hours.

Surprisingly enough, all resistances are very low with the exception of silvered copper in aluminum and in stainless steel with the triethylene-tetramine (IB) cured epoxy.

Table 11-11 shows the high-temperature exposure for all combinations with an additional week at 127C (260F) after all assemblies were encapsulated in an epoxy resin. The previously noted unstable combinations continue to increase in resistance.

Differences due to epoxy systems which were possibly evident in Tables 11-9 and 11-10 become definite in Table 11-11. Comparison of the two high-temperature cure systems indicates that the dicyandiamide-cured epoxy (IIA) is associated with lower resistances than is the aromatic-amine cured epoxy (IA). Comparison of the two low-temperature cure systems indicates that the polyamide-cured epoxy (IIB) is even more definitely associated with lower resistances than is the triethylene-tetramine cured epoxy (IB). Differences associated with the epoxy seem to be primarily confined to those combinations which exhibit metal-to-metal instability; therefore, it seems probable that such differences actually stem from varying ability to inhibit metal instabilities. The surprising fact is that the polyamide-cured epoxy seems to be the best of the lot, even including the two high-temperature cured epoxies.

The data in Table 11-12 show that unfilled-epoxy resins would eventually cause the resistances to increase to infinity from an initial resistance of less than 0.1 ohm. The addition of even 10 percent of a conductive carbon such as acetylene black will keep the resistance within measurable limits, e.g., a maximum of 867 ohms in Table 11-12 after one week at 74C (165F) plus one week at 127C (260F).

Conclusion

On the basis of data for the combinations described here, metal-to-metal incompatibilities are the primary contributors to high resistances in wire assemblies embedded in conductive epoxies. Silver-powder fillers do help to maintain low resistances, and the polyamide curing-agent seems to be the most stable of the four curing-agents evaluated.

Processing in terms of clean surfaces is important, as in any bonding application. Conductive adhesives cause additional problems in that oxidized surfaces which may be acceptable where only strength is required in the bond may be unacceptable where electrical conductivity is needed. Gold- and silver-plating seem to make the cleaning operation less critical, but this could be deceptive, and physical strength might be decreased.

In the study of conductivity of embedded wires it was found that air pockets in the embedment not only reduced the strength of the embedment but increased resistance significantly; therefore, special precautions were taken to insure a complete fill.

Since it was found that commercial adhesives may vary from lot to lot and even within one lot, it is important that assemblies bonded with conductive adhesives be checked 100 percent. But it should be noted that factors affecting reproducibility may not necessarily originate only at the supplier's plant. Time and temperature may cause aggregation of the silver particles and separation from the resin after the materials have passed acceptance testing. Sometimes changes in appearance and consistency may be detected visually, but this is no substitute for control-testing and evaluation of the cured bond.

From the information presented in this chapter it is evident that conductive resins fill a great need but do require some understanding.

REFERENCES

1. Perkins, A. W. *Volume Resistivity and Bond Strength of Some Conductive Adhesives and Sealants.* SC-TM-336-63 (11), Sandia Corporation, December 1963.
2. DeLollis, N. J. "The Use of Adhesives and Sealants in Electronics." *IEEE Transactions*, Vol. PAP-1, No. 3, (December 1965): 4-16.
3. Kilduff, T. J., and Benderly, A. A. "Conductive Adhesive for Electronic Applications." *Electrical Manufacturing*, June 1968: 148-152.
4. Ablestik Adhesive Company, Gardena, California.
5. Keister, Frank Z. "Evaluation of Conductive Adhesives for Microelectronic Applications." *Electro-Technology*, January 1962: 47-52.
6. Data supplied by Bendix Aeronautical Corporation, Kansas City, Mo.

Specification Index

Federal (MMM) and Military (Mil) Specifications	Remarks	QPL (Qualified Products List)
MMM-A-130	Neoprene contact cement for bonding decorative laminates to metal or wood.	No
MMM-A-132	Structural adhesives for use in bonding primary and secondary metallic air-frame parts to withstand exposure to temperatures up to 260C (500F).	Yes
MMM-A-00138	One- and two-adhesive systems for bonding metal to wood.	Yes
Mil-A-5092	Covers three types of rubber cement: I, non-oil resistant; II, oil resistant; III, aromatic fuel resistant for non-structural, general purpose applications.	Yes
Mil-A-14042	Epoxy adhesive (room temperature and intermediate temperature cure) for bonding metals, thermosetting resin laminates, glass, and wood.	No
Mil-A-46028	Adhesive (formulation specification) for bonding polystyrene foam to steel.	No
Mil-A-46091	Brake lining to metal.	No
Mil-A-1154	For bonding non-silicone rubber to steel.	Yes
Mil-C-1219	Water base cement for repair of iron and steel castings.	No
Mil-A-8623	Epoxy adhesives for I, room temperature; II, intermediate temperature; III, high-temperature cured structural bonds to metal, thermosetting plastics, wood, and glass.	Yes
Mil-A-9117	Rubber adhesive for nitrile rubber to metal and for fuel-resistant repair work.	Yes
Mil-C-10523	Fuel-resistant gasket cement.	No
Mil-C-13792	Polyvinyl acetate solution for general purpose bonding of metals.	No
Mil-A-14443	Adhesive for bonding glass to metal, especially in optical systems.	Yes
Mil-S-15204	Thread and joint sealant for temperatures up to 510C(950F).	Yes
Mil-A-18065	Adhesive for securing cork-board insulation to metal.	Yes

Federal (MMM) and Military (Mil) Specifications	Remarks	QPL (Qualified Products List)
Mil-A-22434	The atropic polyester adhesive formulation specification for bonding resin glass laminates to metal, on rough surfaces.	No
Mil-S-22473C	Thread sealant (fourteen grades based on torque strength and viscosity).	No
Mil-A-22895	Polysulfide and rubber base adhesives for bonding name plates to painted and unpainted surfaces.	No
Mil-A-25463	Adhesives (four classes based on temperature resistance for bonding metal sandwich structures to resist temperatures up to 260C (500F).	Yes

Silicone Resins, RTV (Room Temperature Vulcanizing)

Mil-S-23586(WED)	Two-part RTV silicones.	
Mil-A-25457	One- or two-part paste silicone adhesive for bonding silicone rubber to aluminum or to itself.	Yes
Mil-A-46106	One-part RTV silicone rubber adhesive sealants for external seals and for bonding silicone rubber to metal and to itself.	
Mil-D-46838	Two-part RTV silicones.	

Polysulfide Specifications

Mil-S-5817A	Fuel-tank sealant, brush consistency, for protection of synthetic rubber sealants and metal, against oils, fuels, fresh and salt water, corrosion, and weathering.	Yes
Mil-S-7124	A filleting-type sealant for sealing pressurized cabins.	Yes
Mil-S-7126A	Sealant compound for glass and craze-susceptible plastics to metal.	Yes
Mil-S-4383B	Protective coating for synthetic rubber sealants and metal surfaces of integral fuel tanks. A brush consistency, rubber coating compound for the protection of synthetic rubber sealants and metal, against oils, fuels, fresh or salt water, corrosion, and weathering.	Yes
Mil-S-7502C Class A	A brushable sealant for rivets, bolts, and fasteners in fuel tanks and pressurized cabins or cockpits.	Yes
Class B	Fillet sealant for gaps, voids, and seams in fuel tanks and pressurized cabins.	Yes
Mil-S-8516C	For potting and sealing electrical connectors and electric components.	No
Mil-8784 Class A	A brushable sealant for access doors and system components, fuselage doors, and other removable parts. Low adhesion.	Yes

Federal (MMM) and Military (Mil) Specifications	Remarks	QPL (Qualified Products List)
Class B	A fillet sealant for access doors and system components, fuselage doors, and other removable parts. Low adhesion.	Yes
Mil-S-8802D Class A	Brushable material especially developed for use over a temperature range of −65F to +275F with outstanding resistance to jet fuels. Also has excellent resistance to aviation gasoline.	Yes
Class B	Fillet material especially developed for use over a temperature range of −65F to +275F with outstanding resistance to jet fuels. Also has excellent resistance to aviation gasoline.	
Mil-S-11030C	Noncuring polysulfide mastic sealant.	No
Mil-S-11031B	Used as an adhesive sealing compound for bonding metal to metal or glass to metal in optical instruments or fire control instruments.	No
Mil-S-14231B	Brushable material especially developed for faying surface sealing of bolted fuel and oil tanks.	Yes
Mil-C-15705A	A caulking compound for use in seams, joints, and dams formed between metal surfaces to provide watertight and airtight seals.	No

Index

228